Library of Congress Cataloging-in-Publication Data

Burt, Owen H., author. | Burt, Jo, author.
Walter H. Durfee & his clocks / Owen H. & Jo Burt.
  Other titles: Walter H. Durfee and his clocks.
  Includes bibliographical references.
LCCN 2016015113 | ISBN 9781944018023 (pbk.)
Durfee, Walter H., 1857-1939. | Clock and watch
  makers--United States--Biography. | Clock and watch making--United
  States--History. | Clocks and watches--United States--History.
LCC TS544.8.D88 B87 2016 | DDC 681.1/13092 [B] --dc23 LC record available
  at https://lccn.loc.gov/2016015113

© 2017 by National Association of Watch and Clock Collectors, Inc.
All rights reserved. No part of this publication may be stored in a
retrieval system, reproduced, or transmitted in any form by any means,
electronic, mechanical, photocopying, recording, or otherwise,
without written permission from the publisher.

Printed in the United States of America
The National Association of Watch and Clock Collectors, Inc.
Editors: Diana DeLucca, Therese Umerlik, and Christiane Odyniec;
Associate Editors: Freda Conner, Robin Schuldenfrei, & Amy L. Klinedinst
Production Coordinator: Kimberly Hess

Requests to use material from this work should be directed to:
National Association of Watch and Clock Collectors, Inc.
514 Poplar St., Columbia, PA 17512-2130

Founded in 1943, the National Association of Watch and Clock Collectors, Inc. (NAWCC)
is an educational charitable nonprofit member organization whose mission
is to educate the world in the art and science of timekeeping.

See the last page of this book for more information about the NAWCC.

All images in this publication are courtesy of the authors unless otherwise noted.

Table of Contents

Chapter 1 - History & Identification 1

Chapter 2 - Two-Weight Tall Clocks 15

Chapter 3 - Three-Weight Tall Clocks 39

Chapter 4 - Three-Weight Tall Clocks' Movements,
Weights, Dials, Cases, and More 57

Chapter 5 - Curtis Girandole Clocks 73

Chapter 6 - Restoration of a 9 Tube Tall Clock 83

References and Notes 108

About the Authors

Owen Burt was born and raised in Michigan and graduated from Michigan State University, in business management, in 1952. After working in the automotive industry for 35 years, he retired in 1992 to pursue his interest in restoring antique clocks. Owen joined the National Association of Watch and Clock Collectors (NAWCC) in 1970, served on the Board of Directors from 1979 to 1983, was vice president from 1983 to 1985, and was elected president in 1985. Over the past 40 years he has lectured at many National, Regional, and Chapter meetings on "The Welch Patti Clocks" and "Walter Durfee and His Clocks." Owen was made an NAWCC Fellow in 1980 and later received the Star Fellow award. He has been a Life Member of the Association for over 30 years. In 2014 Owen received the NAWCC James W. Gibbs Literary Award for his many articles on clocks that were published in the *NAWCC Bulletin*.

Jo Burt was also born and raised in Michigan and attended Wayne State University. After working 30 years as the assistant to the president of a major hospital in the metro Detroit area, she retired in 2000 to spend more time with her grandchildren. Jo joined the NAWCC in 1972 and served on the Board of Directors from 1987 to 1991. She was awarded the Fellow and Star Fellow for all her contributions to the Association.

They have written many articles for the Association's peer-reviewed journal, the *NAWCC Bulletin*, including "Walter H. Durfee and His Clocks, His Chimes, His Story" (December 1981) and "Seth Thomas 9" Cottage Clocks" (February, May, and June 1998).

They also wrote *NAWCC Bulletin* Supplement No. 12 titled "The Welch, Spring and Company" published by the NAWCC in February 1978.

Their principal interest has been the Welch, Spring "Patti" clocks, Durfee clocks, and late Victorian English bracket clocks.

Preface

The December 1981 issue of the *NAWCC Bulletin* featured an extensive article we wrote on Walter H. Durfee, titled "Walter H. Durfee, His Clocks, His Chimes, His Story." The article gave a brief description of Durfee's life, business, and the clocks he produced and sold from 1880 to 1920. We have spent the past 30 years collecting information about Durfee by speaking with various Durfee clock owners, attending clock auctions featuring Durfee items, examining and recording fake Durfee clocks, and researching old historical records. With this massive amount of new information, we now believe the complete story of Durfee's life, business, and tall clock production can be told.

This book combines the 1981 article with the new information accumulated since that time. The first chapter concentrates on his life and business dealings and provides a brief introduction to the clocks he manufactured. The second through fourth chapters cover in-depth the 2-weight grandfather clocks and the 3-weight chiming hall clocks.

This book compiles six articles published in the *Watch & Clock Bulletin*. Chapters 1 through 4 were published in the January/February 2013, March/April 2013, July/August 2013, and September/October 2013 issues. Chapter 5 was published in the February 2011 issue and Chapter 6 in the March/April 2016 issue.

Acknowledgments

This book could not have been completed without the assistance of NAWCC members, their families, and others who had an interest in Walter H. Durfee and his clocks and who graciously allowed us to photograph and examine their Durfee clocks.

Special thanks to Andrew Apicella, Bruce Austin, Chris Bailey, Ray Beard, Marc Beaulac, Henry Bench, Dana Blackwell, Dan Buffinga, Warner Bundens, Maurice Burns, Herschel Burt, Rick Cantin, Jeremy Ciullo, Jerome and Susan Ciullo, Bruce and Meg Cummings, Joe D'Amico, Jack and Natalie Davis, Delaney's Antique Clocks, Marie DelGatto, Andrew Dervan, Fontaine's Auction, Henry Fried, Tom Harris Auction, David Hoffart, Hope Holdcamper, Daniel Horan, Dennis Jacob, Tom Jepson, David and Ros Jettingoff, James Lapinsky, Eric and Paula Litscher, Dennis Lough, Tran Duy Ly, Lyndhurst of the National Trust, Mirror Lake Inn, Ira Mizrach, Sandy Mrachina, Richard Oliver, Hamilton Pease, Rhode Island School of Design, Robert Schmitt, Harold Smith, South Hampton Antiques, Tom Spittler, Sundial Antiques, John Tanner, George Tardiff, Peter Vander Poel, David Warner, A. Susan Weiser, Stanley Weiss Antiques, Lisa and Edwin Weibrecht, and Diana DeLucca, Therese Umerlik, Christiane Odyniec, Amy Klinedinst, Freda Conner, Robin Schuldenfrei, and Kimberly Hess of the NAWCC Publications staff.

"Grandfather Clock" Expression

The expression "grandfather clock" was a result of a song published by Henry C. Work, in 1876, called "Grandfather's Clock". The song became one of the most popular tunes of its day in America and Britain. By the 1880's the reference to a large standing clock became a grandfather clock here in America, replacing the English term tall clock or longcase that was used through-out Britain. Even today, over a hundred and twenty-five years later many of us know the song that goes,

Lyrics

"My grandfather clock was too large for the shelf,

So it stood ninety years on the floor;

It was taller by half than the old man himself,

Though it weighted not a pennyweight more.

It was brought on the morn of the day that he was born,

And was always his treasure and pride;

But it stopp'd short---never to go again---

When the old man died."

Chorus

"Ninety years without slumbering

(tick, tick, tick, tick),

His life seconds numbering,

(tick, tick, tick, tick),

It stopp'd short---never to go again---

When the old man died."

Walter H. Durfee & His Clocks

Chapter 1

History & Identification

Most Americans are under the impression that the tall clock, or grandfather clock, has been produced and manufactured on a fairly consistent basis since 1700, but that has not been the case. In the United States research has shown that from 1700 to 1775, the majority of all clocks in America were imported from Europe, with the greatest number coming from England. The greatest percentage of clocks made in America were primarily produced in the Philadelphia and Boston areas by craftspeople, who came from Europe, or by individuals who had trained under European clockmakers. It was not until later that the American handcrafted clock industry spread to other metropolitan areas, such as New York, Baltimore, and Newport, RI. Since these clocks were produced by only a small number of clockmakers, it is fair to say that only limited numbers were made during this period of time. It was not until after the Revolutionary War that American clockmakers, such as the Willards, Harland, Mulliken, Chandler, and Terry, started producing complete tall clocks in the United States. The majority of these clocks were similar to the imported clocks, with the cases and brass movements being only an adaptation of the English-style tall clock.

Around 1800 a change took place in the tall clock industry. With the wooden movement replacing the brass handcrafted movement, tall clocks could be produced cheaply and quickly. The wooden movement tall clocks became the accepted standard for the next 35 years. With the introduction of the brass shelf clock with spring-driven movements in the 1830s, the tall clock industry came to a complete standstill. The entire clock industry shifted to producing inexpensive shelf clocks for the masses because there was no demand for tall clocks. During the next 50 years the tall clock industry and market lay dormant.

What does all this have to do with Walter Durfee? In 1885, 50 years after the tall clock industry had died, Walter Durfee (Figure 1.1) rejuvenated the tall clock industry. He started to reproduce and market tall clocks again in

Figure 1.1. Walter H. Durfee (1857-1939).

large numbers, which earned him the title "the father of the modern grandfather clock." More than 100 years ago, he completely changed the clock industry's sales and marketing approach from the economical, inexpensive clocks costing $5 and $10 to expensive clocks retailing for $500 to $700, yet virtually very little has been written about him.

Early Family History

Walter Hill Durfee was born on March 23, 1857, to Elisha A. and Sarah Law (Allen) Durfee in Providence, RI, an area where his Quaker ancestors had settled as early as 1640. Elisha was born on October 17, 1802, in Newport, RI, and his mother, Sarah, was born on August 4, 1830. Elisha was 28 years older than Sarah when they got married on October 9, 1850. She was 20 and he was 48 at that time. Elisha was listed as a trader in the *Providence Directory* as early as 1832. Webster Dictionary describes a "trader," as "a person whose business is buying or selling or barter as a merchant." Exactly what he produced, purchased, or sold is not known. By 1836 his business had grown and Elisha moved his operation to the corner of Spring and Pond streets. He remained at that location as a trader until 1846, when he purchased a grist mill at the corner of Cranston and Dexter streets; the rear of the property faced Durfee Street (named after the Durfee family in prior years).[1.1] In 1860 he had expanded the grist mill business by becoming a kersey manufacturer. Kersey is a coarse ribbed woolen and cotton fabric used for making hoses, work clothes, and uniforms. In 1860 his first known advertising was published in an advertisement in the *Providence Directory*. The ad stated that the factory now had nine sets of machinery and 2,000 cotton spindles for making warps (Figure 1.2). Warps were knitting yarn made in a lengthwise direction. In April 1861, when the Civil War broke out, he obtained a contract with the Union Army to manufacture army blankets. By 1863 he had increased his sales to over $106,000, which included 29,000 army blankets at a sale price of $2.94 each. His income put him in the high 90th percentile for the Providence area. In 1864 his sales income increased to $126,000, and he sold more than 42,000 blankets at $3 each to the Union Army

By this time, Elisha had become quite wealthy and had purchased property and a home at 90 Pond St. in an affluent neighborhood. Most of the people residing in the area had live-in servants, including the Durfee family. Elisha had two successful brothers who also lived on Pond Street. Fred Durfee was a mason and had built the Rhode Island Hospital. Samuel B. Durfee was a prominent builder, and for many years, he was the superintendent of highways in Providence.

Elisha died on February 16, 1865, at the age of 62, just two months before the end of the Civil War. Later, when his estate was settled, we discovered that he not only owned the land under his house but the entire city block, with Pond Street on the north, Winter Street on the west, Spring Street on the east, and Montcalm Street on the south. He also owned the building and property at the northeast corner on Pond and Spring streets just down the street from his mother's home at 90 Pond St. These two properties were willed to his surviving children: Hannah (age 14), Herbert (age 11), Walter (age 7), Elisha (age 5), and Mabel (age 1). The house, the factory, and the bulk of his estate went to Sarah, making her a wealthy widow at the age of 34.

On July 7, 1870, Sarah married widower James A. Wilson, a tailor with a shop in downtown Providence. James Wilson had four small children from his first marriage: Mary A., Florence K., Charles A., and Emma I. Wilson and his children moved into the 90 Pond St. house. This gave Sarah and Mary Dugan, her live-in servant, the responsibility of raising nine children younger than 15 years of age. Although the 1880 US Census shows that James Wilson became the stepfather of the Durfee children, no legal records show he adopted any of the Durfee children.

Life History of Walter H. Durfee

Very little is known about Walter Durfee's early childhood, except he was educated in the Providence public schools and studied to become an architect. After leaving school, he worked as a draftsman at an architectural firm at 45 Westminster St. in downtown Providence. Two years later, he worked as an advanced draftsman on the fourth floor of the Butler Exchange in Providence.

Figure 1.2. Elisha Durfee's 1860 knitting mill ad. *PROVIDENCE DIRECTORY*, 1860.

ELISHA A. DURFEE,

MANUFACTURER OF

KERSEYS.

Cor. of Dexter & Cranston Sts.,

PROVIDENCE, R. I.

☞ Nine Sets Machinery. Also Two Thousand Cotton Spindles for making Warps.

His physical description can best be taken from a 1902 article in the *Providence Journal* about Durfee's run for mayor, which stated:

> Walter Durfee is an extremely handsome man with blue eyes, distinguished silver grey hair, and a matching grey mustache. In a woman, the Durfee type eyes would be called dreamy, soulful, and other lovable terms. In a man, they bespeak simplicity, boyishness and sincerity.

The article also pointed out that he was over 6 feet tall, slender, stood as straight as an arrow, and was always neat and well dressed.

In 1880, when Durfee was only 23 years of age, he gave up being a draftsman at the architectural firm and became a drug clerk for a young druggist named Charles F. Clarke, who was only two years older than Durfee. The drugstore was located at 310 Cranston St. and sold regular pharmaceutical items as well as "medicine man medicinal items," such as Cherry Balsam and Tooth-Ache Cure (Figure 1.3). A short time after working as a drug clerk, he became a partner with Clarke and the business was renamed Clarke & Durfee (Figure 1.4). By mid- or late 1881, the Clarke & Durfee partnership was dissolved and Clarke moved his pharmacy business to his home at 68 Bellevue Ave., a few blocks from the Cranston store, and operated out of his house.

In the meantime, Durfee had set up his antique business at 295 High St. in Providence, at the age of 24. His first ad was published in 1882 in the *Providence Directory* and was listed under antique furniture specializing in English hall clocks (Figure 1.5). For the next two years, he worked on developing a reputation and establishing a good clientele. To accomplish this, Durfee traveled extensively throughout New England and the South looking for English and American antiques, including tall clocks. An example of one of these is the circa 1735 Thomas Brown, of Chester, clock shown in Figure 1.6. Years later Durfee would claim he had purchased and resold over 400 clocks in less than eight years. That calculates to one a week, which is not realistic. Actually, Durfee only sold these antique tall clocks from late 1881 to 1886. By 1886 a majority of his effort was put into manufacturing new clocks. Research has shown that Durfee tended to exaggerate the number of clocks he sold, imported, or manufactured. We would put the number of older tall clocks that went through his hands at less than 100, because we have only verified 15 or 16 of these clocks. Identifying his antique tall clocks is quite easy, because Durfee always attached his Durfee label to the inside surface of the waist door (Figure 1.7).

During these early years of becoming established as an antique dealer, Durfee found local personnel who could restore, repair, and refinish the antiques and tall clocks he was purchasing for resale in his shop. The cabinetmakers he used to restore wood furniture and make tall-case repairs were Rudolph H. Breitenstein and his son,

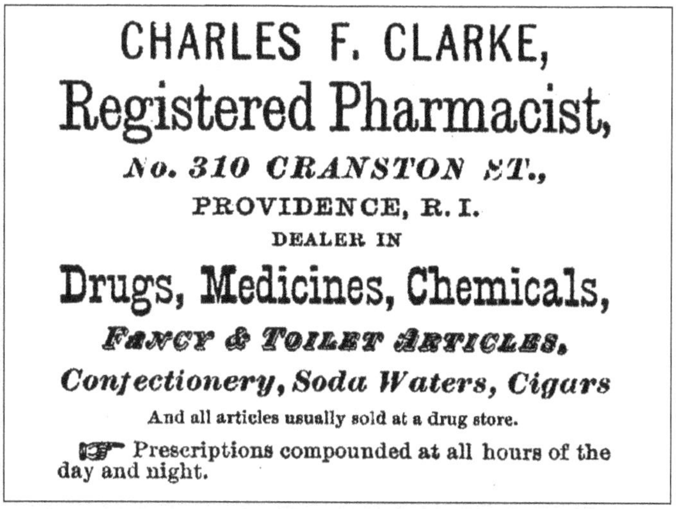

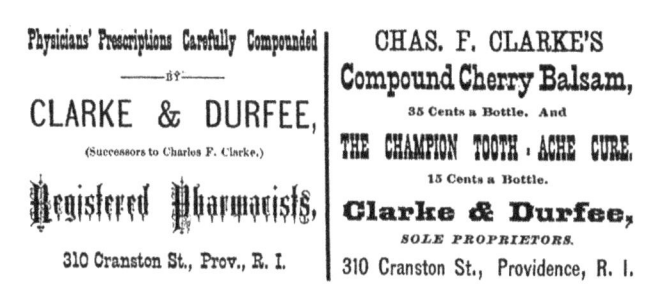

Figure 1.3. Charles Clarke's 1878 pharmacy ad. PROVIDENCE DIRECTORY, 1878.

Figure 1.4. Clarke & Durfee's partnership 1880 ad. PROVIDENCE DIRECTORY, 1880.

Figure 1.5. Durfee's 1882 antique furniture and English hall clock ad. PROVIDENCE DIRECTORY, 1882.

Adolph. They were located less than a mile from his shop at 80 Richmond St. (Figure 1.8). The refinishing work was completed by Rudolph's brother, Otto. The tall clock movements were cleaned and restored by local watch and clock repairer Dexter C. Cheever, initially located at 135 Westminster St. By 1883 he gave his business address as

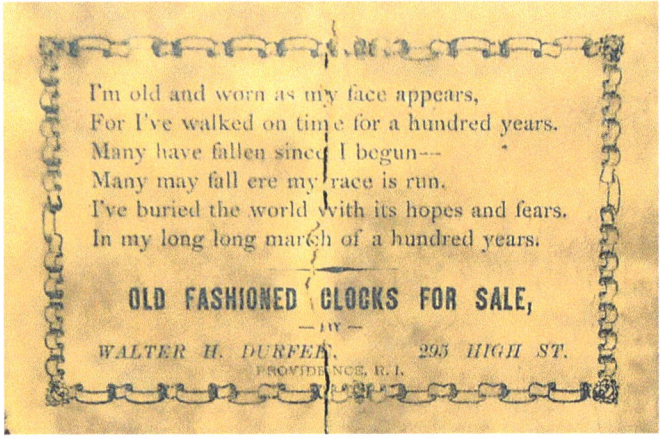

Figure 1.7. Durfee's antique clock label on inside of waist door, used from 1881 to 1886.

Figure 1.8. Providence cabinetmaker R. H. Breitenstein & Son's 1886 ad. PROVIDENCE DIRECTORY, 1886.

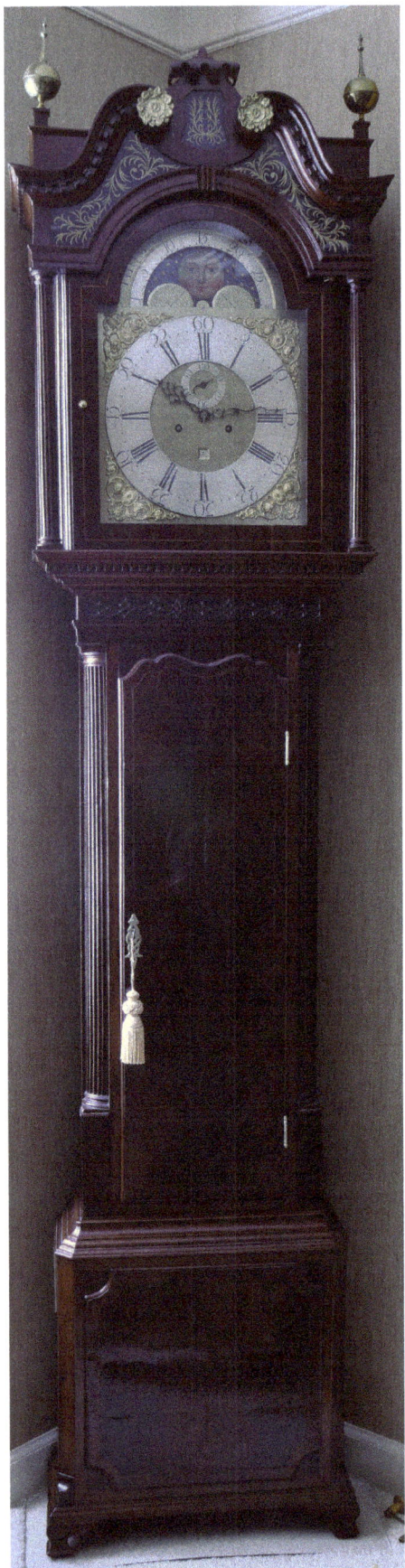

Figure 1.6. Example of an early antique English tall clock sold by Durfee, a Thomas Brown, ca. 1735.

295 High St. It seems that Cheever either was employed by Durfee or was subletting space from Durfee. Or was the young antique dealer, Durfee, subletting space from the more established Cheever? This we may never know for sure. It does appear that Cheever was never a partner in the business with Durfee; the exact relationship has never been determined. Durfee used a second watch and clock repairer for his movement restoration, Jeremiah O. Enches, who worked and lived at 390 High St., just a few doors down the street from Durfee's place of business. It was during these early years in the antique business that Durfee became fascinated with the tall clock and vowed that someday he would manufacture and sell his own clocks.[1,2]

The turning point in Durfee's life came around 1882-1883 when he met an avid antique collector, Charles Leonard Pendleton. Pendleton was born in Westerly, RI, the son of Henry and Rhoda Pendleton, and had attended Phillips Academy, Andover, RI, and Yale University. After graduating from Yale Law School he practiced law in Providence. As Pendleton's interest in antiques grew, his collecting turned into a hobby and an occupation. In his search for fine English and American furniture, Pendleton met Durfee. Sharing a common interest, they soon became friends. Pendleton, besides being an avid antique collector and a practicing attorney, was a man of wealth and a bachelor, and above all, had a weakness for gambling. It is said that Pendleton had made and lost three fortunes in the private card clubs and gambling casinos along the New England coast. Pendleton had three wealthy close personal friends, all of whom were antique collectors and professional gamblers. These men were Richard Canfield of New York and George Palmer and Marsen Perry of New England. Through Pendleton, Durfee became a dealer or agent for all four men.

This relationship with Pendleton grew, and it appears that Pendleton may have started to finance part of Durfee's antique business. Pendleton's search for American and English furniture led him to England several times. In the latter part of 1882, Pendleton, on one of these trips, brought back a few English tallcase clocks that he sold to Durfee. Because old tallcase clocks had been selling quite well for Durfee, we can assume that Pendleton convinced Durfee that he should go to England and purchase them there rather than running around the countryside looking for them. Durfee liked the idea, but he needed someone to help finance his expenses to travel to England. He could not ask Pendleton, because Pendleton had just lost his fortune again while gambling. Therefore, before his May 1883 20-day trip to England, Durfee had to make a special financial partnership agreement with Jeremiah Enches, one of his current movement repairers and an established jeweler and watchmaker, to raise money for the trip. The two men formed a separate clock business Durfee & Enches. It appears that the partnership agreement called for Durfee to purchase new high-grade bracket clocks in London, which were a rage in England at that time, mark them "Durfee & Enches", and resell them through Durfee's antique shop in Providence, RI. Durfee planned to use part of the partnership money to purchase English tallcase clocks.

From all indications, the bracket clocks were purchased from clock company J. C. Jennens & Sons, of London, which had a fine reputation as a bracket clock manufacturer.

Figure 1.9. Durfee & Enches bracket clock, ca. 1883. COURTESY OF PETE VANDER POEL (3).

Figure 1.10. "Durfee & Enches" bracket clock name stamped on movement's backplate.

Figure 1.11. Durfee & Enches bracket clock dial nameplate.

The company was founded around 1860 by John Creed Jennens and Joseph Jennens and remained quite active until the early 1890s. Their shop was located on Great Sutton Street in London, where they manufactured and sold new fine tall clocks and 2- and 3-train fusee bracket clocks. Quite often these Jennens fusee bracket clocks can be found with finely engraved and detailed rear plates, making them highly collectible. Jennens & Sons not only sold directly to the public but also sold their clocks through leading jewelry stores in London, Liverpool, Edinburgh, Glasgow, New York, and Boston. Jennens & Sons made these sales arrangements with Durfee because it gave the clocks another sales outlet. An example of one of the bracket clocks Durfee brought back from England can be seen in Figure 1.9. We do not believe he brought back very many, because after researching Durfee for 35 years, we have only found two. Both of them have the "Durfee & Enches" name on the dial (Figure 1.10) as well as the beautifully engraved back rear plate (Figure 1.11). Durfee also brought back his first new English-cased tall clock (Figure 1.12).[1.3] The tall clock had an inlaid veneered mahogany case stand-

Figure 1.12, left. Durfee & Enches tall clock, ca. 1883. COURTESY OF JEREMY R. CIULLO (3).

Figure 1.13. Durfee & Enches tall clock, 15" dial.

Figure 1.14. Durfee & Enches tall clock's dial nameplate.

Figure 1.15, right. Vernon & Shepperd, Liverpool tall clock, ca. 1795. COURTESY OF DENNIS LOUGH.

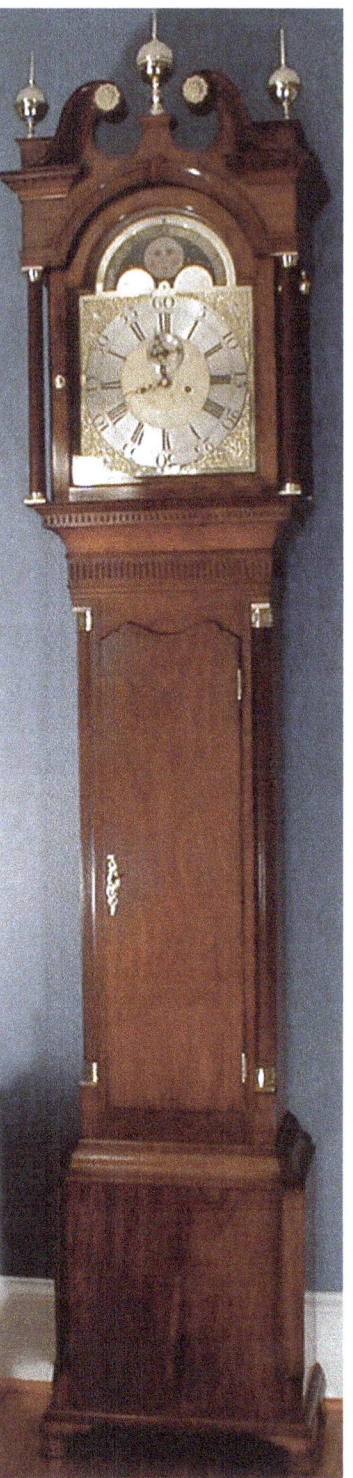

ing over 8' tall, with a traditional rack-and-snail English movement and a large 15" brass dial (Figure 1.13). The dial nameplate reads "Durfee & Enches, Providence", the same as the bracket clocks (Figure 1.14).

Two-Weight Tall Clocks

Besides the purchase of the Durfee & Enches clocks, Durfee also purchased many English tall clocks to resell in his shop to help cover his trip expenses to and from England. Durfee's ship sailed from New York to Liverpool, a perfect place to purchase tall clocks. The Liverpool-Lancashire area cabinetmakers were well known for the distinctive style and very high quality of their tall clock cases. Characteristic features of these clocks are the box top behind the swan neck; the quarter, half, and even full columns on either side of the waist; the shape of the top of the door; the case detailing below the hood and above the waist door; the canted corners on the base; the Chippendale-style feet; and, on more expensive cases, the double columns on either side of the hood.

While in the Lancashire area purchasing tallcase clocks for his shop in Providence, he found one made by Vernon & Shepperd, Liverpool, that he loved (Figure 1.15). This Liverpool clock had all the fine features of a high-quality Lancashire tallcase clock mentioned above. From his experience as a draftsman in an architectural firm, Durfee could see the beauty in this case design and the correct proportions for a perfect reproduction. Using this clock as a model, he found a top-quality cabinetmaker and commissioned him to make a few tall clock cases for him using the Jennens 2-weight movements and dials. When Durfee's reproduction 2-weight tall clocks arrived from England, they were immediately successful with the public.[1.4]

From this 1883 trip, Durfee learned the following:
- He no longer needed a partnership with Enches and that was quickly dissolved.
- He had to find a new financial backer to replace Enches.
- Bracket clocks were not a good investment.
- Reproduction clocks sold better than the antique tall clocks.

Durfee had fulfilled his dream and reintroduced America to the tall clock. Durfee found that his future was in reproduction tall clocks.[1.5]

He realized that going into the clock-manufacturing business required more money than he could afford at that time. It appears that Durfee solved this problem by borrowing money from his mother, who had inherited a considerable amount from the settlement of Durfee's father's estate. With his mother's backing, he made his second trip to England in June 1884. On this trip he intended to coordinate his current and future plans on purchasing 2-weight tall clock components by having the Lancashire cabinetmakers supply the finished cases and having Jennens & Sons supply his movements and dials. Durfee must have purchased or agreed to purchase a large enough quantity of movements from Jennens & Sons for it to agree to stamp his name "Walter H. Durfee, Providence" on the backplate of the movements being supplied to him. The Jennens arrangement may have called for the dials and moondials to be supplied with the movements. Arrangements were made with the different cabinetmakers on the type, style, and quantity of clock cases that he might be purchasing from them in the years to come. Durfee's plan was to purchase movements, cases, and dials in England and have them shipped to Providence, and then he would supply the weights and assemble and test all components before offering them to the public. His advance planning and establishment of credit were accomplished at the age of 27.

The design of Durfee's new clocks combined the styles of Shepperd & Vernon and early American tall clocks (Figure 1.16). The movements made by Jennens & Sons were typical solid brass movements with a rack-and-snail strike. The Jennens movements were factory-stamped "Walter H. Durfee, Providence" on the rear plate (Figure 1.17). Usually, this stamp can be found in the middle of the backplate about 1" from the bottom edge. From 1884 to 1888 most Durfee 2-weight grandfather clocks had solid mahogany cases, a

Figure 1.17. Backplate of Jennens' 2-weight movement with Durfee's stamp.

Figure 1.16. One of Durfee's new clock designs, ca. 1885. COURTESY OF JEREMY R. CIULLO.

solid wooden door, a flat hood glass, a heavy brass dial, a simple calendar, a seconds bit, a gong strike, and a moondial. The moondial had gold-painted stars with a dark blue background.

Because of his new success in the tall clock enterprise, Durfee had to find larger quarters. He rented a building at the rear of 283 High St. for his assembly operation. Because Durfee's clocks were shipped from England disassembled, he needed this extra space to fit the movements, cases, and dials, and test the finished product.

Chiming Hall Clocks

His 1884 plan for importing 2-weight tall clocks worked so well that Durfee did not return to England until 1886. In the spring of 1886, while he was visiting his cabinetmakers, he found that someone was making a prototype case using John Harrington's chiming tubular bells. Harrington, of Coventry, England, had been issued English Patent No. 14,270 on October 28, 1884, that allowed tubular bells to be used in clocks. Harrington had developed and patented a bell chime of such high quality that it could be used in opera houses and theaters. These bell chimes consisted of nickel-plated metal tubes, 1.5" in diameter, which varied in length and were accurately tuned to concert pitch. These tubes produced tones of marvelous sweetness and purity, combined with a vibration that rendered a sound as remote as a distant cathedral spire. The bell chimes had earned the title "the most perfect representation of church chime ever produced." In addition these chimes were awarded a gold medal at Paris in 1885, a bronze medal at the Inventors Fair in 1885, and another gold medal at Liverpool in 1886 (see Figure 2.52 in Chapter 2).

At the time Durfee ran across Harrington's prototype clock case, Harrington had been working with James Jones Elliott, a young clockmaker, who had just formed the firm J. J. Elliott & Co. in London. Elliott was designing and developing a 3-weight chiming movement that could use the Harrington tubes in a tallcase clock. Durfee immediately saw the potential market for the 3-weight chiming hall clock and wanted to add it to his current operation in America.[1,6] It appears that Durfee quickly sought out a private meeting with Harrington to convince him that an agreement between them would benefit both parties. The agreement required Harrington to procure a US patent on his English patent and assign all rights to Walter H. Durfee. In return, Durfee possibly paid Harrington a fee or some type of royalty. Harrington submitted a US patent application on June 11, 1886, that was later accepted and US Patent No. 372,849 was issued. This patent, titled "Chiming Apparatus for Clock", was assigned to Durfee on November 8, 1887.

By the time Durfee received his patent rights in late 1887, Harrington and Elliott had designed a movement, a method of hanging the tubes in the case, and a case large enough to accommodate the long Harrington tubes. The Harrington tubes were suspended in the back of the case on an iron frame directly behind the Elliott movement. This larger movement was capable of striking on eight silvered chime tubes that varied in length from approximately 4' to 5', while the ninth tube, which was 6' long and was used for the hour gong, was suspended from a second iron frame on the right side of the case. With these two working arrangements, one with Harrington and the other with the Elliott & Co., Durfee must have been thrilled, knowing he could enlarge his clock-manufacturing business by adding a 3-weight, quarter-hour strike, chiming tall clock to his current clock line. Prior to his leaving England, he commissioned Elliott and Harrington to proceed with the agreements he had with them to produce the components necessary for him to make the 3-weight chiming clocks in America.

In late November or early December 1887, Durfee received his first chiming hall clock from England. The delay was a result of the US Patent Office not issuing the patent until November 8, 1887. All Durfee's Harrington tubes are stamped with the patent date or dates on them. Therefore, Durfee could not have sold his first chiming tall clock before November 8, 1887. Durfee sold his first 3-weight hall clock for $500. In 1888 a Howard No. 5 banjo clock could be purchased for $20 and a kitchen shelf clock for $5.

Besides having Harrington tubes and the Elliott movement, Durfee's chiming clocks were expensive for the following reasons:

- The cases were made from the finest mahogany, walnut, or oak.
- The cases stood at more than 8' tall (Figure 1.18).
- The cases were heavily detailed with extensive carvings.
- The doors had quarter-inch-thick beveled glass.
- The solid brass dials had raised numerals, silvered chapter rings, and moondials.
- The movement required heavy brass weights and a large brass or mercury pendulum bob.

The chiming hall clock became an instant success for Walter H. Durfee & Co. and with this success came growing pains. Because of the demand for his clocks, Durfee had to enlarge his operation and find larger facilities. He moved his store and manufacturing facility from 295 High St. to 121 Pond St. in Providence. This building, at the corner of Pond and Spring streets, was the same one Walter's father Elisha used while he was a trader from 1836 until 1846, when he purchased the grist mill at 132 Cranston St. This property still belonged to Walter, his brother Elisha P., and his sister Mabel M., all of whom received this property from his estate when he died in 1865. This location at 121 Pond St., by coincidence, happened to be the same building Walter's uncle, Samuel B. Durfee, used to house his horses and oxen when he was superintendent of highways in Providence. Walter H. Durfee & Co. remained at this location until 1920, when

Figure 1.18. Example of a Durfee large 3-weight clock, ca. 1890.
COURTESY OF JEROME AND SUSAN CIULLO.

the city purchased the building for the construction of a new high school. At this location the greatest number of Durfee clocks were produced. The company expanded its business while at this location by adding a second story to the building and an adjacent building. In the 1890s the City of Providence revised Pond Street numbers and assigned 151 Pond St. to the building. The street addresses of 121 Pond St. and 151 Pond St. were used for many years. Around 1905 Durfee used 151 Pond St. in all of his business advertisements and correspondence.

By 1890 Walter Durfee had started a hall clock fad in America and found he could not manufacture the clocks fast enough to meet public demand. To expedite production in England, Durfee decided to spend more time in London. Because of the nationwide demand for Durfee hall clocks, many leading and exclusive jewelry stores requested to be sales agents for Durfee clocks. Durfee had some experience with this agent system while he was dealing with J. C. Jennens & Sons of London to purchase its bracket clocks. Therefore, Durfee made arrangements with exclusive jewelry stores to be his agents and sales outlets for his clocks. To be an agent, the store had to purchase one or two clocks from Durfee's current inventory and retail them at a set price—normally double the wholesale price—with all purchases being F.O.B. Providence, RI. In return Durfee agreed never to sell directly to the public for any price other than the agreed-upon retail price, thereby protecting the agents. As a result of this arrangement, Durfee had more time in England to oversee production. Some of these stores were Tiffany & Co., New York; Theodore B. Starr, New York; Tilden Thurber, Providence, RI; James E. Caldwell, Philadelphia; Spaulding & Co., Chicago; Wright-Kay, Detroit; Schuwanke-Kaston Co., Milwaukee; Boyd Park, Denver; Mermod-Jaccard, St. Louis; J. J. Freeman & Co., Toledo; and Shreve & Co. and Nathan Dohrmann, both of San Francisco.

The Durfee chiming hall clocks produced during the 1890s had either a 5-tube or a 9-tube movement. The 5-tube movement only had the "Westminster" chime, while the 9-tube movement had the "Westminster" chime and the choice of a second distinct melody. Normally, the choice could be made from among the "Whittington," the "Bow Bells," or the "Eight Bell" chimes. Durfee also produced a few less expensive models during this period, which had "Westminster" quarter-hour chime on four accurately tuned cathedral gongs in place of the Harrington tubular bell chime.

By 1897, ten years after selling his first chiming clock, Walter H. Durfee & Co. reported that it had sold over 1,000 chiming clocks at prices ranging from $500 to $700. On an average of $600, this would amount to sales in excess of $600,000 for his chiming hall clocks alone. As mentioned earlier, Durfee tended to exaggerate the number of clocks he sold. We have estimated that he made around 400 grandfather clocks.[1.7] The 3-weight clocks were more popular than the 2-weight clocks and outsold them three to one.

In analyzing Durfee and his clocks, the main reasons for his success are as follows:
- Durfee demanded that his clocks be outstanding, unequaled, and of the highest quality if his company manufactured them.
- Durfee applied his early architectural training to the design of cases that were superior to anything produced ever.
- Most importantly, he held the rights to be sole agent for the Harrington chiming tubes in America, which gave him a monopoly over and control of the market.

Because these chiming hall clocks were in high demand, other clock manufacturers wanted part of this sales market, but Durfee held the rights as the sole agent for the Harrington tubes. As a result, these other companies had to purchase or acquire the tubes or bell chimes directly from Durfee in order to make chiming tubular clocks. These companies included Herschede and

Waltham Clock Cos. Durfee knew his patent rights would expire, so he sold his tubes to other clock manufacturers with special terms and conditions. Durfee believed that if the other manufacturers wanted his tubes, they should also purchase the movements and dials from him. It appears that in many cases he forced them to purchase the entire clock. That is why some early Herschede and Waltham hall clocks have Durfee tubes, movements, and dials.

Breakfast and Stage Chimes

Besides using the Harrington tubes in chiming hall clocks, around 1894 Durfee began using the tubes to produce breakfast and stage chimes for use in houses, theaters, opera houses, and churches. These breakfast calls, or chimes, consisted of two, three, or four tubular bells mounted to a mahogany bracket hung on the wall. An example of these breakfast call chimes can be seen in Figure 1.19. The stage chimes were a set of tubular bells that gave a perfect imitation of distant cathedral chimes. The stage chimes could be put in portable form, for ease of transport and so were used by many leading theatrical companies. The breakfast calls and stage sets were awarded a gold medal at the Mechanics Fair in Boston in the fall of 1895. Durfee and Harrington received world recognition for these theater and opera chimes.

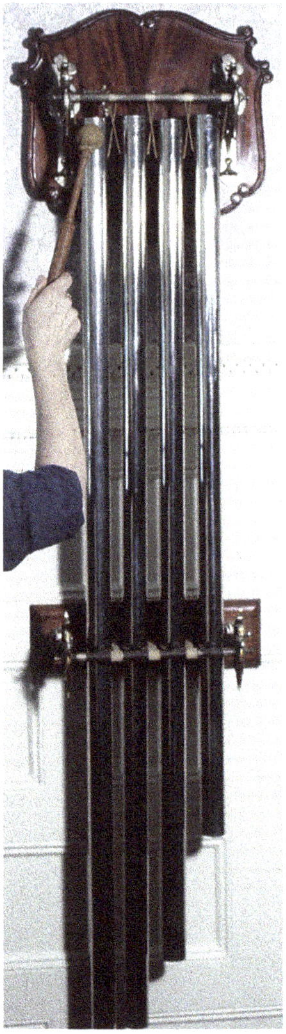

Figure 1.19. Durfee's 4-tube breakfast chimes.
COURTESY OF TED BURLEIGH.

Setback by Bawo & Dotter

In 1902 Durfee had his first major setback in business. In 1900 a clockmaker named Bawo, from Brooklyn, NY, started to use chiming tubes that were not Durfee tubes. Durfee insisted that this was an infringement of his patent rights and sued. The case was tried in the US Federal Court of the Southern District of New York. The judge ruled, on October 4, 1902, that, "the right of the public, the use of what has been disclosed by the prior British patent, the monopoly having expired, should not be curtailed. A decree may be entered dismissing the bill." Durfee lost the lawsuit and the judge's decision terminated the Walter H. Durfee & Co.'s monopoly.

By 1905 Frank Herschede had designed his own chiming tube and had it patented. Within a few years, Waltham Clock Co. had tubes, and by 1908 it appears that every US clockmaker had gotten into the chiming hall clock business. With so many manufacturers producing chiming hall clocks, competition became competitive. By 1910 prices started to drop and so did quality, but Durfee refused to lower either. Although Durfee was disappointed by this change of events, he knew that by this time he had placed approximately 300 Durfee chiming hall clocks in prominent houses, universities, and businesses throughout the United States.

Hall Clock Replacement Business

With the change in the hall clock industry, Durfee set his time and talent in another direction. He noticed that Waltham had started to make a reproduction of the early Willard banjo clock, and he could see a definite sales market for the banjo. Durfee was not a stranger to the banjo clock; while in the antique business, he had become an admirer of Lemuel Curtis and his clocks, and later an authority. The December 1923 issue of *Antiques* magazine published an article by Walter H. Durfee titled "The Clocks of Lemuel Curtis."

Curtis Girandole[1.8]

In November 1907 Durfee reproduced his first banjo clock, a copy of Curtis's girandole (Figure 1.20). Durfee was not the type to want to copy exactly; he wanted to improve on everything, even Curtis's girandole. Durfee's girandole, in many people's opinion, is the finest reproduction ever produced. The mahogany case is 44-1/2" long and is stamped, "Walter H. Durfee, Providence" and dated November 1907. It has a Waltham movement with sapphire inserts on the pallets; brass bezels and brass balls; the dial is fired porcelain enamel on copper; and the reverse glasses were painted by an unknown artist. The painting on the lower glass represents *Paul Revere's Ride*, with the Old North Church in the background. Durfee added his special touch by attaching painted porcelain inserts at the top and bottom corners of the two side arms. Durfee's girandole is nearly a perfect copy of Curtis's girandole that was owned by the late Mrs. Benjamin Peckham, except for these added porcelain inserts. The acanthus bracket carvings, the wooden eagle, the gold leafing, and the reverse paintings are outstanding and breathtaking and could even have earned a smile from Curtis himself.

Banjo Clocks

From 1908 through 1918 Durfee concentrated on producing banjo clocks. He limited his Willard banjo line and only manufactured four different styles. The four styles were the standard Willard banjo in a mahogany finish (Figure 1.21), the gold-leafed presentation model, the Curtis-style lyre in a 42" long case (Figure 1.22), and a short lyre in a 32" case. These Willard-style cases were

Figure 1.20. Durfee's girandole clock, ca. 1920.

Figure 1.21. Durfee's Willard-style banjo clock in mahogany case, ca. 1915.

Figure 1.22. Durfee's lyre banjo clock in mahogany case, ca. 1915.
COURTESY OF GEORGE TARDIFF (3).

made of the finest mahogany, beautifully hand finished, with polished brass side arms and solid brass bezels. The top of the case was decorated with a hand carved wooden eagle, a cast eagle, or an acorn finial, but in most cases they were done in old gold.

Willard Striking Banjo

In his earlier years when he was selling antiques and antique clocks, Durfee ran across a Roxbury striking banjo movement someone had put in a Howard clock case. Durfee knew this striking banjo movement had to have been made by either Simon or Aaron Willard.[1,9] Durfee kept the movement, waiting until the proper time to use it. In 1912 he reproduced a Simon Willard banjo using this Roxbury striking movement (Figure 1.23). In producing this striking banjo, Durfee put to use some of his finest skills and workmanship. At the time, Durfee intended to keep this clock, but after it was completed, for unknown reasons, the clock was sold to a private party in New York. In a letter dated October 25, 1916, to the new owner, Durfee explained in detail that the clock was designed by Durfee in the Simon Willard style. To make certain that the clock was never passed on as a Simon Willard, Durfee stamped "Simon Willard" on the inside of the case. This is the only banjo clock we have found that was made by Durfee where he didn't sign the dial "Walter Durfee".

Elisha "Chester" Durfee

Except for the first few years in business, Durfee was always the sole owner and manager of Walter H. Durfee & Co. In 1914, when Walter Durfee was 54 years old, he

Walter H. Durfee—Chapter 1 • 11

took into his business his brother Elisha P. Durfee's only son, Elisha Chester Durfee, who was 24 and had just graduated from Brown University with a bachelor's degree in history. For the first couple of years, Elisha was a clock winder for Walter H. Durfee & Co. This position developed around the turn of the century in many metropolitan areas with affluent clock owners. For a service charge or fee the clock winder would wind the owner's clocks once a week. The fee would be determined by the distance the clock winder had to travel and the number of clocks they had to wind. The clock winder, on their weekly trips, would check the clocks for service or adjustment. If they found that a clock needed repair or cleaning, they would advise the customer and take the clock back to the shop for service and repairs. Later, because Durfee was getting on in years and wanted to leave the business to someone, he taught the clock-making and repair business to Elisha.

From 1917 through 1930 it appears that Durfee's clock-manufacturing business came to a standstill. We believe that the clock industry had changed; the demand for quality was gone; and the clock market was now crowded with cheaper and inferior clocks. Durfee, now in his sixties, refused to lower his high standards to compete in this new market. From all indications, Walter H. Durfee & Co. only produced a few banjo clocks during this time and the majority of Walter's and Elisha's time was spent repairing and servicing clocks for people in the Providence area. Occasionally, they acted as agents, dealers, or appraisers to those who were interested in purchasing or disposing of antique clocks.

Wallace Nutting

During these later years Walter Durfee became a close friend and advisor to famous antique expert Wallace Nutting. Nutting even gave credit to Durfee as being his authority in Nutting's book *The Clock Book*, published in 1924. Each man admired the other's ability and skill, and at one time, the two made a professional trade: Durfee traded one of his chiming clocks to Nutting in exchange for a matched set of eight Nutting-made dining room chairs. In 1928, when Durfee, at the age of 71, became a grandfather to his only grandchild, Rosemond Downs, Nutting personally made a youth chair for the baby.

Figure 1.23. Durfee's Simon Willard reproduction striking banjo clock, ca. 1915. ANONYMOUS.

Family Life

Walter Durfee lived and worked primarily in an area less than two miles in diameter despite traveling extensively for business. He lived with his mother at 90 Pond St., a few blocks from the 295 High St. store, until his marriage to Jessie Segar in 1884. After they were married, Durfee purchased a house at 5 Lester St., just three blocks from his mother's house. Unfortunately, Jessie Durfee passed away in 1893. Two years later, in 1895, Walter married Florence Wilson and moved to 5 Franklin St., five blocks from the manufacturing plant at 121 Pond St. They had two daughters, Louise and Isabel, while living on Franklin Street. They moved a short distance away to Elmstead, then to Mawney, then to President Street. While at President Street, Louise married a Mr. Downs and moved out. Durfee's wife, Florence, died in 1923. By that time, the shop had relocated to 270 Washington St. His unmarried daughter, Isabel, was a teacher at the high school, and she remained at home, 599 Angell St., with Walter until he died August 4, 1939, at the age of 82. Walter was buried in the family plot at Swan Pointe Cemetery in Providence, RI.

Antique Dealer

Earlier we briefly mentioned that Durfee was primarily in the antique business when he opened his shop in 1881. Durfee never left the antique business, and it remained part of his daily operations and income up to 1930. It did not take Durfee long to realize that money could be made in high-quality English and American antiques. With that in mind he scoured the East Coast and the South looking for quality antique furniture. This led to his meeting Charles Pendleton, a major antique collector and dealer. Pendleton taught Durfee the skills and tricks of being a successful antique dealer. With Durfee's architectural training, his love for high-quality items, and his desire to learn, it did not take him long to become an expert in the antique business. On his many trips to England he purchased quality antiques, as well as clocks, to bring back. It appears that in many cases he acted as a broker or middleman for other antique dealers. By 1900 he was considered an authority on the subject and had many affluent collectors seeking his advice on items they owned or were anticipating purchasing. During these years Durfee and Pendleton became very close personal friends. In 1904, when Pendleton was dying of cancer, he made Durfee the executor of his estate, which included Pendleton's antique collection. Pendleton, a bachelor with no heirs, had willed his estate to the Rhode Island School of Design in Providence, with the understanding that it would build a fireproof "house of Georgian architecture" to display his antique collection. The gift included some 1,400 objects, including 130 pieces of English and American furniture, and English and Continental pottery and porcelain. In addition, he left two lots on Benefit Street, adjacent to Brown University, for the school to construct the Pendleton House Museum. It took Durfee more than two years, as executor, to settle the estate.

Figure 1.24. Durfee patent date stamped on Harrington tubes.
COURTESY OF JEROME AND SUSAN CIULLO.

Political Involvement

Durfee got involved with politics in 1899 and served in the Rhode Island State House of Representatives until 1902. He ran unsuccessfully for mayor of Providence in 1902 as a Republican. The main cause of his defeat was his refusal to spend any of his own money to campaign. Most people in Providence knew very little about him and referred to him as "the clockmaker candidate." A 1902 article in the *Providence Journal* about Durfee running for mayor published this comment about him: "As already suggested, there are clocks and clocks. The kind that Mr. Durfee puts together is tall hall clock variety—tall in stature, tall in price."

Identifying Durfee Clocks

So often people, resale shops, and auction houses believe they have a Durfee chiming hall clock because the tubes have the "Walter H. Durfee" name and patent dates stamped on them. They may not know that Durfee sold his tubes to other clock manufacturers. Because of this, we want to explain how to identify a true Durfee. Our research has shown that if a tall clock was truly manufactured and sold by Walter H. Durfee & Co., the clock must meet all of the following criteria:

On a tubular hall clock, all tubes must be stamped "Walter H. Durfee" and must show at least one of his four patents or dates (Figure 1.24). The earliest tubes had one patent dated 1887; the next group had patent dates of 1887 and 1888; the third group had patent dates of 1887, 1888, and 1892; and the last group had dates of 1887, 1888, 1892, and 1896.

The back of the dial plate and the moondial will be stamped "Walter H. Durfee" at the top of a 1" circle, the bottom of the circle reads "Providence," and the circle center reads "R.I." (Figure 1.25).

Most importantly, the movement must be signed "Walter H. Durfee."[1.10] In some cases "Walter H. Durfee, Providence," was factory stamped on the lower surface of the backplate by the movement manufacturer during the movement's construction. This method of identification was primarily on the 2-weight movements. On

Figure 1.25. Durfee's circle signature stamp, used on back of moon and dial plates.

Figure 1.26. "Walter H. Durfee" stamped on back edge of movement rear plate. COURTESY OF RAY BEARD.

the 3-weight movements purchased from J. J. Elliott & Co. of London, Durfee chose not to have the movement stamped "Walter H. Durfee" by the manufacturer. He did this for a very special reason. Being an enterprising businessman when he obtained the US patent rights to the Harrington chime tubes, he could foresee the possibility of selling the tubes and other clock components to other US clockmakers at a markup. When other clock manufacturers wanted to purchase the Harrington tubes from Durfee, he normally sold them a package that consisted of the Harrington tubes, the dial assembly, weights, pendulum, and the J. J. Elliott movement. Because Durfee was only selling the component package, he left the casing entirely up to the clock manufacturer. These component sales to other clock manufacturers, such as Herschede and Waltham, became a lucrative part of the Walter H. Durfee & Co. sales. Durfee was extremely proud of his clock-manufacturing enterprise and his reputation in the clock industry. He did not want others to capitalize on his name or his reputation. Because Durfee had no control over the case styles of the component buyers, he devised a method of identifying Durfee-made clocks by Walter H. Durfee & Co. All movements used in his clocks were signed "Walter H. Durfee." This was accomplished by having the movement maker, other than J. J. Elliott Co., stamp "Walter H. Durfee" on the movement (Figure 1.17), and in the case of the J. J. Elliot movements, by having Walter H. Durfee & Co. handstamp "Walter H. Durfee" on the top edge of the backplate (Figure 1.26). All authentic Durfee grandfather clocks will be signed in one of these two methods.

Other helpful, but not required, ways to identify a Durfee grandfather clock are as follows:
- "Walter H. Durfee" stamped in a 1" circle on the seat board (Figure 1.27)
- "Walter H. Durfee" on the nameplate of the dial
- "Walter H. Durfee" stamped on the top face of the dial between the hemispheres (Figure 1.28).

As a young man, Walter Durfee studied to be an architect, which influenced many of his case designs. Durfee cases are well balanced in design and carry some characteristics or lines of early tall clocks. All Durfee tall clocks had feet, whether they were Chippendale, carved, clawed, or reverse-ogee style.

Summary

To some, he has been considered a contemporary clockmaker who produced high-quality banjo and tall clocks. It is true that Durfee never made his movements, dials, cases, or other components of the clocks. Joseph Martines's article in the December 1977 *NAWCC Bulletin* on contemporary clockmakers refers to Walter Durfee as an orchestrator.[1.11] Webster's Ninth Collegiate Dictionary defines an orchestrator as "one who composes or arranges for an orchestra." Durfee was more than an orchestrator; he was a symphony conductor. Like a great conductor, who put together different sounds and produced a beautiful symphony, Durfee designed, composed, and brought together the finest components available and with a conductor's touch produced a symphony of clocks.

Figure 1.27. Durfee's circle signature stamped on movement's seat board.
COURTESY OF RAY BEARD.

Figure 1.28. "Walter H. Durfee" stamped on face of dial between hemispheres.
COURTESY OF GEORGE TARDIFF.

Walter H. Durfee & His Clocks

Chapter 2

Two-Weight Tall Clocks

Figure 2.1. Pattern 1, Durfee & Enches tall clock, ca. 1883. COURTESY OF JEREMY R. CIULLO (3).

Walter Durfee produced and sold 2-weight tall clocks from 1883 until 1915. However, no known catalogs were published by Durfee or Walter H. Durfee & Co. during that time, so we have taken the liberty of assigning clock pattern numbers to each currently known case style. In the early 1890s, Frank Herschede of Cincinnati, OH, published a trade catalog of the clocks he was currently selling in his jewelry store in Cincinnati. Herschede, an agent for Durfee's chiming hall clocks, included in the catalog only those Durfee clocks he chose to sell. For his own reference purposes, Herschede assigned a pattern number to each clock in his catalog. Research has shown that Herschede did not use any Durfee model or pattern numbers, but used his own numbering system. In 1994, when Tran Duy Ly published *Long Case Clocks and Standing Regulators Part I: Machine Made Clocks*,[2.1] Herschede's catalog was incorporated in the Durfee section of the book. Since that time, Tran's book is a top reference for identifying clocks Walter H. Durfee & Co. manufactured. However, a number of Durfee clocks are not shown in Tran's book, and we question whether some clocks in his book are authentic Durfee clocks. We attempt to clarify Durfee production by illustrating and describing in detail the known clocks Walter H. Durfee & Co. manufactured.[2.2]

Our Numbering System

Herschede used only nine of the numbers between 2 and 20 for the 3-weight chiming hall clocks, leaving ten unused numbers. When cataloging the 2-weight tall clocks, he used only four numbers: 41 through 44. For future identification purposes, we have assigned Herschede's unused pattern numbers to previously unidentified Durfee 2- and 3-weight clocks. Herschede's unused number 1 was given to Durfee's first reproduction grandfather clock, labeled as "Durfee & Enches", and manufactured in 1883. Following Herschede's numbering system, the 3-weight chiming hall clocks' numbers will range from

Figure 2.2. Pattern 1, marquetry inlay above waist door.

Figure 2.3. Pattern 1, base and claw feet.

number 2 through 25. The numbers for the remaining 2-weight tall clocks will run from number 30 through 44. In both cases, we have tried to assign the numbers in chronological order, based on when we believe new models were first introduced. Not all numbers will be used, so future discovered clocks can be assigned to them.

Durfee & Enches (Spring 1883) – Pattern 1

In May 1883, on his trip to England, Durfee brought back his first reproduction tall clock, Pattern 1 (Figure 2.1).[2.3] When using the phrase "reproduction tall case clock," we are implying that Durfee purchased a new case from a cabinetmaker, had it fitted with a new movement and dial, and assigned his name to the clock. The Pattern 1 mahogany case was made in the Liverpool area and styled after early Lancashire tallcase clocks. The clock is 8'7" tall, 24" wide, and 12" deep. The marquetry case has an elaborate decorative pattern above and below the waist door (Figure 2.2). There is a 3/8" checkered inlaid box pattern on the face of the base and a 1/8" inlaid banding on the hood door and the waist door. The case has eight full, fluted columns with brass Corinthian capitals and bases. The hood has four columns, two in the front and two in the rear, which are only found on the more expensive and decorative Lancashire tall clocks. The waist area is 18" wide, which is 2"-3" wider than the 2-weight clock cases that he would eventually produce. The case has heavy carved claw feet as shown in Figure 2.3. Durfee used claw feet, like these, in his later more expensive clocks. There is a Durfee label on the inside of the waist door (Figure 2.4). Durfee used this label when he first started selling antique tall clocks in 1882 but stopped using it by 1885.

The S. Harlow, Derbyshire, England, 8-day brass movement, made between 1875 and 1882, has a rack-and-snail strike, striking on a cast "bell-metal" bell (Figure 2.5).[2.4] This is the only known movement Durfee used with a cast bell; in all of his future 2-weight tall clocks, he used a flat wire gong strike. The movement does not strike the half hour, and the motion for the calendar and the moondials operates off the snail wheel. In a few of his early clocks, Durfee added an iron shelf bracket fastened to the back of the case with the base of the bracket attached to the movement seat board (Figure 2.6). The movement has cast-iron weights, 1-1/2"-diameter pulleys, a wooden pendulum rod, and a 5"-diameter brass pendulum bob.

The clock has a 15" broken arch brass dial, with a moondial, second-bit chapter ring, a crested shaped calendar-wheel aperture, brass cherub spandrels, and a raised, silvered, engraved chapter ring. The dial nameplate reads "Durfee & Enches, Providence" (Figure 2.7).

Durfee & Enches Variant (Mid 1883) – Pattern 1-V1

In the late summer of 1883 Durfee had his second known clock shipped from England (Figure 2.8). The case appears to have been made by the same cabinetmaker as

Pattern 1.[2,5] The clock is a sister of his Durfee & Enches 2-weight tall clock, except for a few variations in the case, movement, and dial. We refer to this clock as Durfee's Pattern 1-V1, with the V1 standing for variant number 1. The clock has the same overall dimensions, finials, swan necks with brass trim, four fluted columns on the hood, marquetry below the hood, and the same base front and claw feet. The Pattern 1-V1 case replaced the fluted columns on the waist and base with small fluting in the surface. The case has a shorter waist door with a much larger inlay on the lower portion of the waist, and its base has canted corners (Figure 2.9). Durfee discovered that the clock looked better with a 13" dial than a 15" dial. To use the 13" dial, Durfee had the cabinetmaker install a recessed 1"-wide filler panel around the dial opening to eliminate the void. The dial spandrels were upgraded to more detailed cherubs and Durfee would use these on future clock dials (Figure 2.10). The dial's nameplate was

Figure 2.4. Pattern 1, Durfee's antique clock label on inside of waist door.

Figure 2.6. Pattern 1, angle brackets (see arrows) holding movement seat board down and hour gong mounted to backboard.

Figure 2.5. Pattern 1, S. Harlow English movement used in Durfee & Enches tall clock.

Figure 2.7, right. Pattern 1, Durfee & Enches dial nameplate. COURTESY OF JEREMY R. CIULLO (4).

Figure 2.8. Pattern 1-V1, Durfee tall clock, ca. 1883. COURTESY OF DENNIS L. JACOB (3).

changed to "Walter H. Durfee, Providence," replacing "Durfee & Enches" on Pattern 1 (Figure 2.11). The movement is exactly the same as the previous movement, except a motion wheel was added in the upper right corner of the front plate, for moving the moondial (Figure 2.12). Durfee had a wire gong installed that gave a softer sound to the hour strike and replaced the cast bell-metal bell used in the previous clock.

Durfee's First Designed Case (1884) – Pattern 30

While in the Liverpool area in 1883 Durfee had purchased a Lancashire 1760s antique tall clock, made by Vernon & Shepperd, of Liverpool, for re-sale in his shop in Providence (Figure 2.13).[2.6] Durfee was impressed with its styling and eye appeal; it had all of the finer Lancashire's tall clock case characteristic features, for which the area was famous. He decided to design and build his first completely new Durfee tallcase clock. He commissioned a cabinetmaker to construct a case using the Vernon & Shepperd clock as a model, while incorporating his own personal ideas and specifications. The end result was Pattern 30 (Figure 2.14). Pattern 30 is 101" tall, 20-1/2" wide, and 11" deep. The Vernon & Shepperd's and Pattern 30's overall dimensions are basically the same, except Pattern 30 is 2" shorter.

Both cases have the following features in common:
- Box top behind the swan necks
- Brass rosettes on the swan necks
- Four fluted columns with brass Doric capitals and bases on the hood
- Mahogany case
- A 1/8" inlay banding around the hood door
- A 15"-wide waist
- Two fluted quarter-round columns on the waist with brass capitals and bases
- Two case bases of the same size with canted corners.

Figure 2.9. Pattern 1-V1, large marquetry inlay below waist door and above base.

Figure 2.10. Pattern 1-V1, 13" dial.

Figure 2.11. Pattern 1-V1, dial nameplate and calendar opening.
COURTESY OF DENNIS L. JACOB (2).

Figure 2.12, below. Pattern 1-V1, S. Harlow movement with added moon motion wheel (see arrow).

Figure 2.13. Vernon & Shepperd's Lancashire tall clock, ca. 1760.
COURTESY OF DENNIS LOUGH.

Walter H. Durfee—Chapter 2 • 19

Durfee always tried to improve upon the design of any product he was making. In this case, Durfee made the following changes:
- Reduced the waist door size by 1" to make it 10" wide
- Redesigned the top of the waist door from the Lancashire wave to a semicircle
- Added a chain flower inlay on the waist above the top of the door
- Added a 1/2" box pattern inlay in the base face, making the base similar to the Durfee & Enches clock
- Replaced the Chippendale feet with claw feet (Figure 2.15).

Pattern 30 has a J. C. Jennens & Sons 2-weight movement with cast-iron weights and a brass pendulum bob. The 13" arch top brass dial has a silvered, raised, chapter ring and second bit dial. The dial's four brass spandrels and calendar aperture are similar to those used on the Pattern 1 dial. The dial nameplate reads "Walter H. Durfee, Providence" and is the same shape as the one on the Pattern 1-V1 dial (Figure 2.16). The moondial has a cobalt blue background with gold stars, which became his standard for years to come.

Case Design Experimental Years 1884-1885

Besides working with the Pattern 30 cabinetmaker, it appears that Durfee was experimenting with other casemakers to find the best tallcase design for his US market. We have found only five different models, Patterns 30-34, that Durfee produced during this period. We have described in detail Pattern 30, which was copied after the Vernon & Shepperd clock. The four remaining clocks, Patterns 31-34, vary in their case design and eye appeal. We can assume that he only produced a small number, because only one of each model has surfaced over the past 30 years. We are not positive which model or style was made first, but we believe they were all produced between late 1884 through 1885.

Pattern 31 – 1885

Pattern 31 has a mahogany veneered case with 1/2" inlay banding above the hood door, the outer edges of the waist door, and a large square in the lower base (Figure 2.17). The four corners of the rectangular waist door banding and the lower base square have marquetry fan inlays. The waist door veneer has a beautiful pattern.

Figure 2.14. Pattern 30 tall clock, ca. 1884. ANONYMOUS (3).

Figure 2.15. Pattern 30, lower base and claw feet.

Figure 2.16. Pattern 30, dial and Durfee nameplate.

Figure 2.17. Pattern 31 tall clock, ca. 1885. COURTESY OF ERIC AND PAULA LITSCHER (3).

Figure 2.18. Pattern 31, silvered dial.

Figure 2.19, below. Pattern 31, side view of hood showing glass sides (see arrow).

Figure 2.20. Pattern 32 tall clock, ca. 1885. COURTESY OF DR. WARNER BUNDENS.

The lower base veneering in the square is quarter-cut to match panels. The dial has a silvered, raised chapter ring and a silver hand-engraved center with Durfee's arrow nameplate (Figure 2.18). The hood has two solid wood columns with wooden capitals and bases. A very unique feature of Pattern 31 is the hood with glass side panels (Figure 2.19). The waist and base have canted corners, and the base has Chippendale feet.

Pattern 32 – 1885

Pattern 32 has a solid mahogany case that is 102" tall, 20" wide, and 11" deep (Figure 2.20). The hood has wooden columns and capitals while the swan necks' rosettes are hand carved. The waist has quartered, fluted columns and wooden capitals. The base has a recessed face and Chippendale feet. A brass plate on the waist door dates the clock to the 1885 period. The plate reads, "Presented

Figure 2.21. Pattern 32, Jennens & Sons movement.
COURTESY OF DR. WARNER BUNDENS (3).

Figure 2.23. Pattern 32, dial and jeweler's nameplate, J. E. Caldwell, Philadelphia.

Pattern 33 – 1885

Pattern 33 has a mahogany case that is 93" tall, 25" wide, and 12-1/2" deep (Figure 2.24). This model was featured in Durfee's first known clock advertisement in the 1885 *Providence Directory* (Figure 2.25). The hood has four unusual barley twisted columns with wooden capitals and bases. The top of the hood has a large 7" high ornately carved finial. The hood and waist doors have string brass inlays on the outer edges of their frame. The waist is 16-1/2" wide, making it 1" to 1-1/2" wider than the standard 15" waist used on most of Durfee's 2-weight tall clocks. The waist door is 11-1/2" wide, making the waist appear wider than it is. The door size and design are similar to the waist he used in Pattern 1 and Pattern 1-V1 (See Figures 2.1 and 2.8). The waist has fluted wooden columns with wooden capitals and bases. The 24"-wide base has four recessed panels and stylized feet.

Figure 2.22. Pattern 32, Walter H. Durfee's factory stamp on movement rear plate.

to the Library of the Physicians of Philadelphia by S. Weir Mitchell 1886." Dr. S. Weir Mitchell was the president of the College of Physicians from 1886 to 1889 and 1892 to 1895. It has the J. S. Jennens & Sons 2-weight brass movement with the Durfee stamp on the back (Figures 2.21 and 2.22).[2,7] It has a 13" brass dial with brass cherub spandrels and a moondial with the blue background and gold stars (Figure 2.23). The dial nameplate design is the new one used in Patterns 1-V1 and 31.

It appears that Durfee used an early version of the Jennens & Sons 2-weight 8-day time-and-strike brass movement with moon phase and calendar. The 13" silvered brass dial has cherub spandrels, a raised Roman numeral chapter ring, a raised second bit chapter ring, a calendar opening, and a beautifully engraved floral dial center. The dial is similar to the dial in Figure 2.18. The moondial has the painted blue background with gold stars pattern. Durfee used the same arrow-shaped nameplate he used in Patterns 31 and 32. Pattern 33 also has cast-iron weights, brass pendulum bob, and brass pulleys. Between 1883 and 1884 Durfee sometimes stamped his name on the small semicircle between the two spheres of the dial (Figure 2.26).

Pattern 34 – 1885

Pattern 34 has a beautiful marquetry case that can be classified as one of a kind (Figure 2.27). Durfee only used wood inlay on a few of his early clocks and in limited amounts. This case is an exception to the rule and shows the work of a quality casemaker. Whether Durfee was

Figure 2.24. Pattern 33 tall clock, ca. 1885.
COURTESY OF R. O. SCHMITT FINE ARTS, LLC.

Figure 2.25. Pattern 33, 1885 ad. *PROVIDENCE DIRECTORY,* 1885.

Figure 2.26. Pattern 33, Durfee's name stamped between hemispheres on dial. ANONYMOUS.

Walter H. Durfee—Chapter 2 • 23

testing the market to see if there was any interest in heavy marquetry or whether he had a customer who had requested this style, we may never know. The clock is 100" tall, 22-1/2" wide, and 10-1/2" deep. The hood has four twisted columns, a carved urn finial, wheat stalk inlay above the door, and beveled glass in place of the usual flat glass he used on previous clocks (Figure 2.28). The waist has twisted columns with carved capitals, recessed carvings above the door, and a leaf and flower pattern inlay on the door face (Figure 2.29). The base of the case has a recessed face with a beautiful inlay of leaves and finely detailed carved claw feet.

Although this clock has a 3-weight movement, we have placed it in the 2-weight tall clock category because of its narrow 15" waist size, which is standard for the 2-weight cases. In Chapters 3 and 4, when we cover the 3-weight hall clocks, it will be explained that the 2-weight case needed a wider 18" waist to accommodate the Harrington tubes. The movement and dial were made by W. F. Evans & Sons, Soho Clock Co., Handsworth, Birmingham, England (see Figure 2.30). The firm was established in 1805, but it did not export clocks or movements to the United States until the 1880s. Durfee is said to have imported some W. F. Evans & Sons clocks or possibly just movements and stamped his name on them. It strongly appears that in this situation the case and movement were made by W. F. Evans & Sons.[2.8]

The W. F. Evans & Sons movement is quite large because it chimes "Eight Bells" on eight nickel-plated bells and the "Westminster" chimes on four wire gongs, while the hour is struck on a larger wire gong (Figure 2.31). The brass dial has silvered raised chapter and second rings plus three silvered raised dials in the arch area (Figure 2.32). The top ring is for the silent and chime lever, the right ring controls the chime selection for either the "Westminster Chimes" or "Chimes on Eight Bells," whereas the left ring is for the fine adjustment of

Figure 2.27. Pattern 34 tall clock, ca. 1885. COURTESY OF STANLEY WEISS (3).

Figure 2.28. Pattern 34, hood.

Figure 2.29. Pattern 34, waist door.

Figure 2.30, above. W. F. Evans & Sons ad. *JEWELERS' CIRCULAR AND HOROLOGICAL REVIEW*, 1888.

Figure 2.31, right. Pattern 34, W. F. Evans chiming movement. COURTESY OF STANLEY WEISS (3).

Figure 2.32. Pattern 34, W. F. Evans dial.

Figure 2.33, right. Pattern 34, dial nameplate and chime chapter rings.

Figure 2.34. Pattern 34, close-up of Durfee's nameplate. COURTESY OF STANLEY WEISS (2).

Figure 2.35, right. Pattern 34, cylindrical lead pendulum bob.

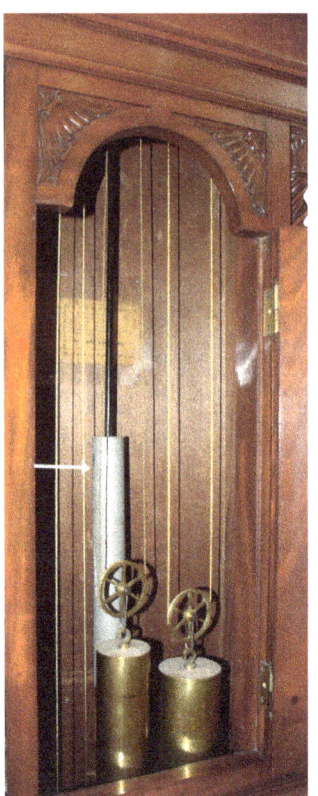

the pendulum for fast or slow (Figure 2.33). Below the chime-silent ring is a silver nameplate that reads "Walter H. Durfee Providence" (Figure 2.34).

The pendulum does not have a standard brass bob. The movement was made by W. F. Evans & Sons that used a heavy cylindrical lead bob that is 1-5/8" in diameter and 12-1/2" long with a wooden rod (Figure 2.35).

Pattern 35 – 1886 through 1888

By 1886 it appears that Durfee was through experimenting with different case designs and settled on the Pattern 35 design, which he would use for the next three years (Figure 2.36). The solid mahogany case is 101" tall, 23" wide, and 11" deep with a waist width of 15". Durfee would use these four basic case dimensions on all of his 2-weight clocks for the next 29 years.

The Pattern 35 case has the following characteristics:
- Swan necks with brass rosette trim (Figure 2.36A)
- Six fluted full columns with gold-gilded Corinthian capitals and bases
- Flat glass in the hood door, not beveled
- Waist door made of solid mahogany and finished on the inside and the outside
- Broken arch at the top of the waist door
- Chippendale feet on the base.

Pattern 35's costly and unique feature in its case design is its 1/16" inlaid brass strips that encompass the hood door glass, the open area below the hood, the area above the waist door, the waist door, the area below the waist door, and the face of the base (Figure 2.37). In all, the inlaid brass stripping is more than 400" long, which includes the 11"-diameter circle in the base and the hyperbola curve shape in the waist door.

Pattern 35 has the Jennens & Sons 2-weight brass movement with cast-iron weights, 2-1/2" diameter six-spoke brass pulleys and a 7" brass pendulum bob (Figure 2.38). The 13" brass dial has brass cherub spandrels and a silvered raised chapter ring, second bit dial, and Durfee nameplate (Figure 2.39). The moondial has the blue background with gold stars.

Figure 2.36. Pattern 35 tall clock, ca. 1886-1888. COURTESY OF JEREMY R. CIULLO (5).

Figure 2.36A. Pattern 35, hood and waist with brass inlay strips.

Figure 2.37. Pattern 35, base with brass inlay strips and Chippendale feet.

Figure 2.38. Pattern 35, cast-iron weights and brass 6-spoke pulleys.

Figure 2.39. Pattern 35, dial and Durfee's nameplate.

Figure 2.40. Pattern 35-V1 tall clock, with Evans & Sons 3-weight movement, ca. 1886. COURTESY OF TOM HARRIS AUCTIONS (2).

Figure 2.41. Pattern 35-V1, hood and dial.

Figure 2.42. Pattern 36 tall clock, glass waist door, ca. 1888-1892. COURTESY OF GEORGE TARDIFF.

Figure 2.43. Pattern 37 tall clock, ca. 1889-1890. COURTESY OF GEORGE TARDIFF (2).

Pattern 35 Variant (1886) Pattern 35-V1

Durfee made one variant of Pattern 35, which we have labeled as Pattern 35-V1 (Figures 2.40 and 2.41). The clock case is identical to Pattern 35, except it has claw feet, not Chippendale feet. It is listed as a variant because it has the 3-train W. F. Evans & Sons movement and dial, which we described in Pattern 34, in place of the standard Jennens & Sons 2-weight movement. The dial nameplate is similar to the one shown in Figure 2.32 and reads, "Walter H. Durfee, Providence, R.I."

Pattern 36 – 1888 through 1892

Pattern 36 is a duplicate of Pattern 35, except for the waist door (Figure 2.42). Starting in 1888 Durfee's clocks changed from a solid wood waist door to a waist door with a beveled glass. From this time on, all of Durfee's 2-weight and 3-weight clocks had a beveled-glass hood and waist door glass.

Pattern 37 – 1889 through 1890

Pattern 37 (Figure 2.43) is similar to Pattern 36, except for the following:
- Twisted columns replaced the six full-fluted columns[2.9] and the right-side twist mirrored the left-side twist
- Flat front base was altered to include a recessed insert panel (Figure 2.44)
- Hard maple veneer strips replaced the brass inlay strips.

Figure 2.44. Pattern 37, base with marquetry inlay and barley twisted columns as well as claw feet.

The twisted columns used in Pattern 37 became a standard feature on all future Durfee 2-weight clocks. Patterns 37 and 38 can be classified as "transition clocks," because their design is a step above the major design change in Pattern 35, but they are a step below the final changes made in Pattern 42.

Pattern 38 – Late 1889 through 1890

The Pattern 38 tall clock case (Figure 2.45) is identical to Pattern 37, except that the base's front insert panel reverted to a flat surface and the veneer inlay was completely eliminated. The case came in either solid mahogany or oak and is 99" high, 24" wide, and 14" deep. As of the date of this publication only two Pattern 38 cases—one of solid mahogany and one of oak—are known to exist. The provenance of the Pattern 38 mahogany case shows that Durfee referred to this model as No. 16 1/2 and informed the new owner that the clock had a London movement of the highest quality, striking the hour and half hour on a wire gong. It appears that Durfee purchased movements for the Pattern 38 cases from a supplier other than Jennens. These new movements (Figure 2.45A) are 5-1/2" wide and 7" high. They could have been supplied by J. J. Elliott & Co., because they are similar to the one shown in Tran Duy Ly's book, *Longcase Clocks and Standing Regulators*.[2.10] The Elliott and Pattern 38 movements required a sounding board for the gong, which is behind and below the movement. The Pattern 38 movements are handstamped "Walter H. Durfee" on the top edge of the backplate. These 2-weight movements also required a special lever, mounted to the back on the dial plate, to move the moondial (Figure 2.45B). This moondial lever arrangement is the same basic design used by J. J. Elliott & Co. to activate the moondial in its 3-weight movements (see Figure 1.25 in Chapter 1). This is the only known time Durfee used these 2-weight movements. Because of the rarity of the Pattern 38 cases and movements, this model has become desirable.

Figure 2.45. Pattern 38 tall clock, mahogany case, ca. 1889-1890.
COURTESY OF FONTAINE AUCTIONS (3).

Figure 2.45A. Pattern 38, London-made 2-weight, movement, which strikes the hour and half hour on a wire gong.

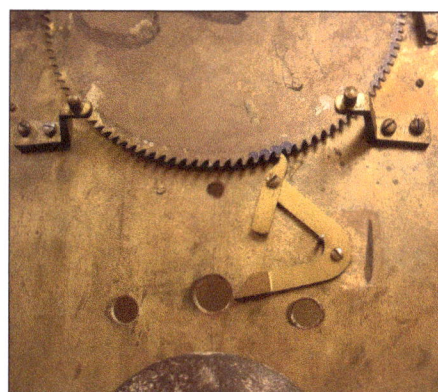

Figure 2.45B. Pattern 38, lever for moving moondial, mounted on back of dial plate.

Pattern 38-V1 - Late 1889 through 1890

Pattern 38-V1 (Figures 2.46 and 2.46A) is identical to Pattern 38, except it has a 3-weight movement. It appears that the Eight Bell chiming movement was supplied by J. J. Elliott & Co., London, and can play either the Westminster chime or eight bell chime on the quarter hours (Figure 2.47). Because J. J. Elliott did not stamp "Walter Durfee & Co." on the movement, Durfee handstamped his name on the top edge on the rear plate to identify that his company made the clock (Figure 2.48). Durfee had started stamping his name on the movement in this location as early as 1885. It appears that all movements coming unsigned from Elliott, including the 3-weight movement, would be signed in this method. The brass dial does not have a moondial but has the control levers for the chiming (Figure 2.49). The dial nameplate is engraved "Walter H. Durfee".

Pattern 42 – 1890 through 1915

From 1883 to 1890, Durfee "tinkered" with his 2-weight case design, trying to find one that was perfect for the current clock market. In about 1890 Durfee made his last 2-weight design change and produced Pattern 42 (Figures 2.50 and 2.51). Durfee used this basic design or variants of this model in either mahogany or oak cases exclusively from 1890 through 1915.

The number 42 was assigned to this model by Herschede Clock Co. when it published its 1895 trade catalog. Historical records show that Durfee's advertising fliers called it Pattern 16 up until that time (Figure 2.51). Because most people refer to it as Pattern 42, we thought it best to use this name.

Pattern 42 was first introduced to the public in 1890 in a Durfee advertisement published in the *Providence Directory* (Figure 2.52). Pattern 42 has the following standard features that make it exceptional, starting from the top of the clock and working down:

- Three large brass finials
- Two swan necks at the top of the hood with two ornate, brass, rosette stampings (Figure 2.53)
- Six twisted columns with gilded brass Corinthian capitals and bases

Figure 2.46. Pattern 38-V1 tall clock, chiming, oak, ca. 1889. COURTESY OF GEORGE TARDIFF (3).

Figure 2.46A. Pattern 38-V1, base of clock in Figure 2.46.

Figure 2.47. Pattern 38-V1, J. J. Elliott 8-bell chiming movement.

Figure 2.48. Pattern 38-V1, "Walter H. Durfee" stamped on top edge of rear plate (see arrow). COURTESY OF GEORGE TARDIFF (2).

Figure 2.49. Pattern 38-V1, arch dial without moon.

- Beveled hood door glass
- Brass dial with a moondial, a second bit, and a calendar (Figure 2.54)
- Heavy high-grade brass English movement
- A 15"-wide waist
- Carved semicircle fan on the top of the waist door
- Bevel glass waist door
- Brass weights and pulleys
- Beautifully carved sunburst pattern in the base with carved quarter fans in the outer corners (Figure 2.55).

In mid-1890 Durfee made his last movement change. His unknown movement source supplied him with a movement not of a standard rectangular shape, but one with legs that extended to the sides (Figures 2.56 and 2.57). Durfee identified it as a Durfee movement by stamping "Walter H. Durfee" on the top of the rear plate (Figure 2.58).

There are exceptions to the large brass finials and the twisted columns with gilded brass Corinthian capitals and bases listed above on one clock we have seen. This Pattern 42 had carved wooden finials and carved capitals and bases. Most Pattern 42 clocks have claw feet but a few have Chippendale feet (Figure 2.59).

Pattern 42-V1 – 1890-1

Pattern 42-V1 is the only variant we have seen (Figure 2.60). It could be a one-of-a-kind special order. The oak case is heavily carved on the six side panels as well as the top surfaces of the base front and sides (Figure 2.61). The moondial has the blue background with gold stars, placing its age circa 1890-1891, because Durfee stopped using the blue background around this time.

Figure 2.50. Pattern 42 tall clock, ca. 1890-1915. COURTESY OF JEROME AND SUSAN CIULLO.

Figure 2.51. Pattern 42, Durfee's personal ad for his No. 16 clock.

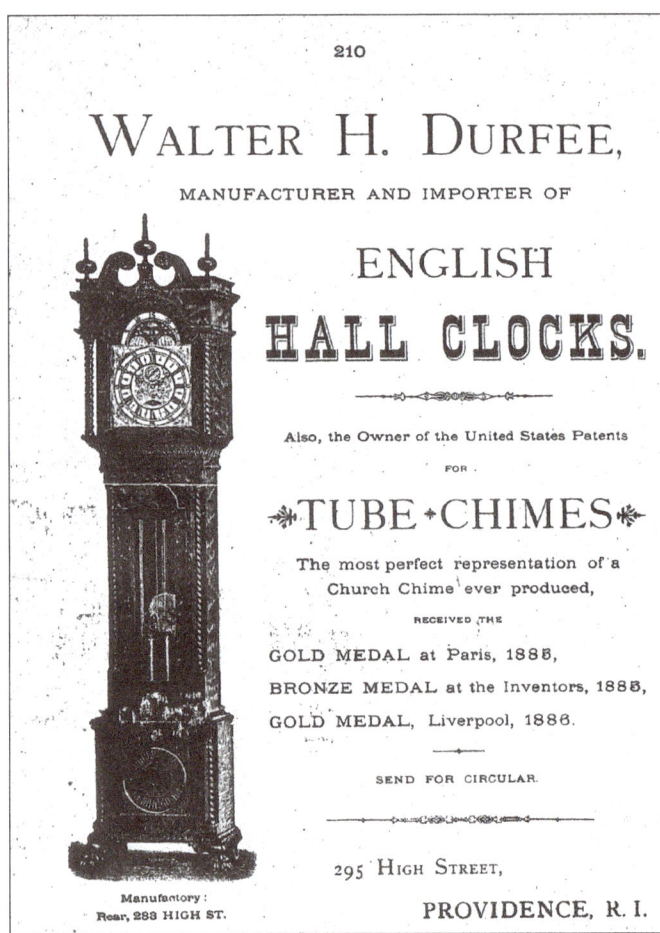

Figure 2.53. Pattern 42, finials, swan necks, rosettes, etc. COURTESY OF JEROME AND SUSAN CIULLO.

Figure 2.52. Durfee's Pattern 42 tall clock ad. PROVIDENCE DIRECTORY, 1890.

Figure 2.54, right. Pattern 42, dial. COURTESY OF RAY BEARD.

Figure 2.55. Pattern 42, base with carved sunray design. COURTESY OF JEROME AND SUSAN CIULLO.

Figure 2.56. Pattern 42, front view of movement, ca. 1895. COURTESY OF RAY BEARD (2).

Figure 2.57. Pattern 42, rear view of movement, ca. 1895.

Figure 2.59. Pattern 42, Durfee's name stamped on top edge of rear movement plate (see arrow). COURTESY OF RAY BEARD.

Figure 2.61, left. Pattern 42-V1, carved base.

Figure 2.58. Pattern 42, Chippendale feet. COURTESY OF SOUTH HAMPTON ANTIQUES.

Figure 2.60. Pattern 42-V1 tall clock, heavily carved, ca. 1890-1891. COURTESY OF GEORGE TARDIFF (2).

Pattern 42 Presentation Clock – 1897

Over the years we have seen many Durfee clocks that had an engraved presentation plate attached to the case. Some clocks were given as an award, some as a birthday present, some as an anniversary present, and some as a wedding gift. The presentation clock that we have enjoyed the most is one that was given in 1897. A Pattern 42 was sold to a groom's seven best friends, who gave it to the bride and groom as a wedding present. Because the groom and his friends were yachtsmen, who loved the sea, his friends added seashells and brass trim to the case, so when the groom wound the clock it would remind him of the sea and his friends (Figures 2.62 and 2.63).

Herschede Pattern 41 and 43 – 1895

Both of these clocks are shown in the 1895 Frank Herschede trade catalog and Tran Duy Ly's *Longcase Clocks and Standing Regulators*.[2.11] Because we have never seen nor known of anyone who owns or has seen either one, we can only provide our personal opinions on these two clocks.

By examining the hand drawing in Herschede's trade catalog, we found that Pattern 41 (Figure 2.64) has the few following characteristics that could make it a Durfee clock:

- Top of the hood is designed like a Napoleon hat, similar to the top of Pattern 5.
- Rosettes below the top of the hood are similar to those on Pattern 4.
- Engraved panel above the waist door is similar to the one on Pattern 20.
- Six-spoke pulleys are the same as those on many Durfee 2-weight clocks.
- Dial, hands, and the dial nameplate are identical to those shown in the Herschede catalog for Pattern 42.
- Hood button latch is one used on many 2-weight clocks prior to 1898.

We believe this could have been a Durfee-designed clock; whether one was ever made is up to speculation.

The same cannot be said of Pattern 43 (Figure 2.65), for there is not one characteristic that appears to make it a Durfee design or clock. The dial, pulleys, feet, and door latches are not items that Durfee ever used. Our conclusion is that Pattern 43 is a Herschede design and had nothing to do with Durfee.

Pattern 44 – 1895

One Pattern 44 clock (Figure 2.66) has surfaced in the past 30 years, and it is definitely a Herschede clock (Figure 2.67). The dial nameplate reads "Frank Herschede, Cincinnati" (Figure 2.68). The movement is not one that was ever used in a Durfee clock. The case design does not have any Durfee features. Therefore, we believe that Pattern 43 and 44, both 2-weight tall clocks, were designed, manufactured, and cataloged by Herschede. It appears that Herschede was only an agent for all Durfee 3-weight tube clocks, and Pattern 41 and Pattern 42 are in the 2-weight category.

Pattern Summary

Durfee only made 11 patterns of 2-weight clocks from 1883 to 1915. The vast majority were manufactured between 1888 and 1903. The two most popular were Patterns 35 and 42, which account for about 80 percent of all Durfee 2-weight clocks produced. Patterns 35 through 38 account for approximately 25 percent of all Durfee 2-weight clocks made and sold by Walter H. Durfee & Co., while Pattern 42 accounts for about 55 percent. We believe the rarest Durfee clock is Pattern 1, his very first clock. Patterns 30 through 34 are desirable and collectible due to the small number made.

Figure 2.62. Pattern 42-V1, presentation clock with seashells on case, ca. 1897. COURTESY OF GEORGE TARDIFF (2).

Figure 2.63. Pattern 42-VI, presentation clock base with brass shell and trim.

PATTERN No. 41
MOUNTED WITH EITHER MOVEMENT.
FOUR GONG AND EIGHT BELL CHIME.
HALF-HOUR STRIKE, First Quality.
FOUR GONG CHIME
HALF-HOUR STRIKE, Second Quality
{Photo courtesy of Cincinnati Historical Society}

PATTERN No. 43
MOUNTED WITH EITHER MOVEMENT.
FOUR GONG AND EIGHT BELL CHIME.
HALF-HOUR STRIKE, First Quality.
FOUR GONG CHIME
HALF-HOUR STRIKE, Second Quality
{Photo courtesy of Cincinnati Historical Society}

PATTERN No. 44
MOUNTED WITH EITHER MOVEMENT.
FOUR GONG AND EIGHT BELL CHIME.
HALF-HOUR STRIKE, First Quality.
FOUR GONG CHIME
HALF-HOUR STRIKE, Second Quality
{Photo courtesy of Cincinnati Historical Society}

Figure 2.64. Herschede's cataloged Pattern 41. This image is Figure 354 on page 152 in Tran Duy Ly's *Longcase Clocks and Standing Regulators* from arlingtonbooks.com. COURTESY OF ARLINGTON BOOKS.

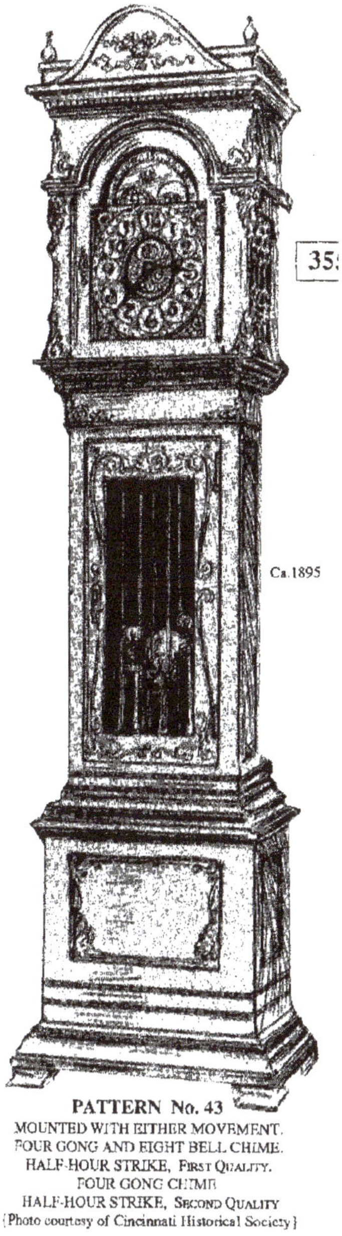

Figure 2.65. Herschede's cataloged Pattern 43. This image is Figure 355 on page 153 in Tran Duy Ly's *Longcase Clocks and Standing Regulators* from arlingtonbooks.com. COURTESY OF ARLINGTON BOOKS.

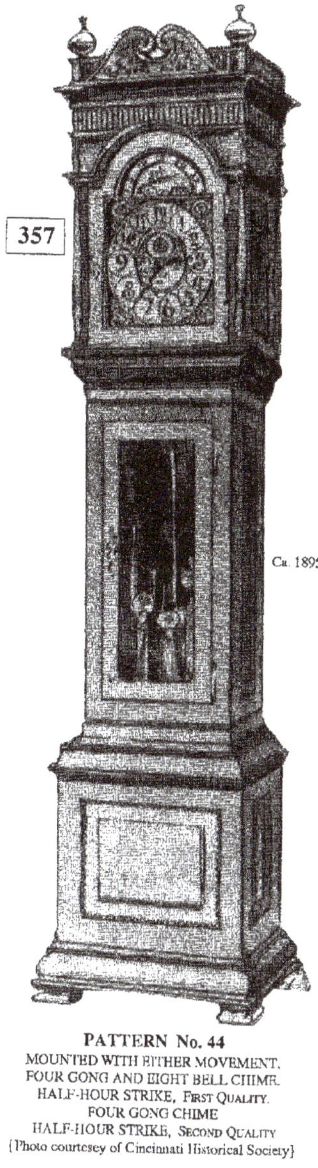

Figure 2.66. Herschede's cataloged Pattern 44. This image is Figure 357 on page 154 in Tran Duy Ly's *Longcase Clocks and Standing Regulators* from arlingtonbooks.com. COURTESY OF ARLINGTON BOOKS.

Figure 2.67. Herschede's actual Pattern 44. ANONYMOUS.

Figure 2.68, left. Dial of clock in Figure 2.67, with "Frank Herschede Cincinnati" nameplate. ANONYMOUS.

Durfee 2-Weight Clocks – Facts & Information

- All 2-weight cases were imported from England up to 1889-1890, after which they were all made in the United States.
- The cases were made from either mahogany or oak.
- All cases came with a high-quality, handrubbed finish.
- Cases had either carved claw feet or Chippendale feet.
- Starting with Pattern 35, all future Durfee 2-weight cases had swan necks with rosettes, six columns with gilded capitals and bases, and three brass finials.
- The six columns were always full columns and were either fluted or barley twisted.
- Up until 1890 the hood had button latches.
- Starting in 1888 all waist and hood doors had beveled glass.
- The flat brass waist door escutcheon (see Figures 2.1 and 2.29) was used until 1889, after which the escutcheon became a cast brass lion head (see Figure 2.45).
- The dials were all broken arch.
- The moondial had a cobalt blue background with gold stars from 1883 until early 1892 when Durfee started using landscape and nautical backgrounds.
- The backs of all moondials and brass dial plates have "Walter H. Durfee, R.I." stamped in a 1-1/2" circle, except the dials used in Patterns 38 and 38-V1.
- Durfee started using his famous double-arrow dial nameplate (Figure 2.11) in late 1883. He continued using it until 1915 on 80 percent of his 2-weight clocks.
- The majority of the double-arrow nameplates are signed "Walter H. Durfee, Providence," not the name of a retail jeweler.
- In the early years, Durfee also stamped his name on the dial in the small semicircle between the two dial spheres below the moondial (Figure 2-26).
- The movements were factory stamped (Figure 2-22) or handstamped on the top edge of the rear plate (Figure 2.48).
- The movements were made in England, and J. C. Jennens & Sons was the major supplier until the mid-1890s, when Durfee used a new source. Because the new movements were unsigned by the factory, it is quite difficult to determine who the new supplier was.
- Durfee used a few 3-train striking movements made by W. F. Evans & Sons until 1888.
- Durfee used a flat wire gong with all standard 2-weight movements, except for Pattern 1, which used a cast bell-metal bell.
- The flat wire gong with a cast-iron base was normally mounted to the backboard, except for Pattern 38.
- In the one Pattern 42 case, the wire gong was replaced with a 72" Harrington tube mounted on the right-hand side of the case.
- The English sources for the 13" dials have never been determined, but a few of the 2-weight dials have "Jennens & Sons" stamped on the brass dial plate under the 6 o'clock of the chapter ring.
- Dials came with either Roman or Arabic numerals.
- Most dials have the cherub spandrels.
- All 2-weight dials have a second bit and a calendar opening.
- Durfee did not use painted dials until 1900 and then he used only a few.
- All early Pattern clocks with solid waist doors used cast-iron weights painted black.
- Durfee purchased cast-iron and brass weights in England until 1888, after which he had all weights made and assembled in the Providence area.
- Durfee used the 2-1/2"-diameter 5- and 6-spoke pulleys as early as 1885 and on and off until 1905.
- When Durfee wasn't using the spoke pulleys, he used the standard 1-1/2" pulley.
- The waist door hinges were standard brass hinges until around 1885-1886 when some clocks had fancy engraved hinges.

General Operation Summary

In 1882, when Durfee started his antique business at 295 High St., Providence, the shop was nothing more than a small store, about 500-600 square feet. When he started selling tall clocks a few years later, he needed more space. In 1886 Durfee rented a building at 283 High St. at the rear of his store. He used this building as his workshop for restoring antiques and antique English hall clocks as well as an assembly area for his reproduction 2-weight tall clocks he was importing from England. As business improved, he again needed more space. In 1889 he moved to 121 Pond St., the property and building owned by the Durfee family. This increased his operation to around 2,000-2,500 square feet. His operation remained there until 1923 when the City of Providence purchased the land for a school. Durfee moved to a much smaller store front at 270 Washington St.

It appears that Walter H. Durfee & Co. was mostly a one-person operation with Durfee doing practically everything.[2.12] As was previously stated in Chapter 1, from 1883 through 1915, Durfee only manufactured approximately 400 tallcase clocks, of which 125 were the

2-weight clocks. Around 25 of the 400 clocks were produced from 1883 to 1887, and they were primarily the 2-weight tall clocks. Similarly, he made approximately 25 clocks between 1904 and 1915. This low volume occurred when Durfee lost his patent rights to be the sole supplier of the Harrington chime tubes in the United States in the *Bawo vs. Durfee* lawsuit, in 1902. After 1904 Durfee's tall clock business came to a complete halt.

From 1888 through 1903 Durfee was most productive and produced 350 tallcase clocks, averaging 24 clocks a year or 2 per month. His method of operation was quite simple:[2.13]

- He had the movements, pendulums, and dials shipped to him from England.
- After 1888, he had the cases and weights made in the United States and then shipped to his shop in Providence.
- He then fitted the movement to the case.
- Last, he tested the clock for a few weeks.

All of this work could be done by one person in a relatively short time, giving him plenty of time to run his antique business and market his clocks. It appears that Durfee purchased his movements, dials, and the brass weight shells in lots of 10 or 20. The same cannot be said of the cases, for it was difficult to predict which case or cases the public was going to want or order. Durfee must have made commitments to his casemakers on the number he would purchase, but style would be determined by public demand.

Selling Prices & Manufacturing Cost

By using 1890-1900 trade catalogs, we could determine the manufacturing cost and selling prices for Durfee's 2-weight tall clocks. Examples are as follows:

- Brass moondials cost $15-$21
- A 1/2-hour strike movements cost $21
- Mahogany cases cost $30-$45
- Weights and miscellaneous items cost $12-$15.

Low-end clocks would cost $78, and Durfee would add his markup and sell them to the retailers for $150. The retailer would double their purchase price and sell the clock for $300. The high-end clocks would cost $102, and Durfee would add his markup and sell them wholesale to the retailer for $200. The retailer then would sell to the public for $400. Clocks were never put on consignment to the retailer, payment was made in advance, and all shipments were FOB Providence, R.I.

For comparison purposes, the Durfee 2-weight clocks sold for $300-$400, while his competitors were selling a much lower quality clock for $150-$200. Yet Durfee outsold them. As far as appreciation, Durfee's competitor's $200 2-weight tall clocks resell for $1,000-$1,500 in pristine condition as of the date of this publication. Durfee's 2-weight tall clocks resell for ten times that amount. Durfee's insistence on high-quality, grace, and beauty in the clocks he manufactured paid off not only in the beginning but also in later years. Because of the special efforts Durfee put into his products, the public as well as collectors have sought out his clocks.

Walter H. Durfee & His Clocks

Chapter 3

Three-Weight Chiming Tall Clocks

During the past 30 years we have uncovered a vast amount of new material about Walter H. Durfee's 3-weight tall clocks, so much so that two chapters are necessary to present it. This chapter covers the 3-weight tall clock cases that Durfee used from 1887 until around 1912. Chapter 4 examines the movements, dials, tubes, pendulums, weights, finials, feet, hardware, and sales agents Walter H. Durfee & Co. used in the 3-weight tall clocks.

Identifying a True Durfee Clock

Over the years several 3-weight tall clocks have been misrepresented as Durfee clocks by dealers, auction houses, antique stores, and private individuals seeking to gain a better price for them. Just because the tubes are stamped "Walter H. Durfee" does not mean the clock was made, produced, or manufactured by Walter H. Durfee & Co. Durfee controlled the US patent rights to the Harrington chime tubes and profited from the overall production of chiming tall clocks. He required American clock manufacturers, such as Herschede, Waltham, Tobey, and Horner, to purchase the tubes directly from him as well as his package of component parts at a markup. This package consisted of tubes, an unsigned J. J. Elliott & Co. movement, a dial, the necessary weights, and a pendulum. According to Elisha Durfee, the component package cost almost as much as a new clock.[3.1] Durfee's financial arrangement kept his competition from undercutting his prices.

Our research has shown that in determining if a chiming tall clock was truly manufactured and sold by Walter H. Durfee & Co., it must meet all of the following criteria:

- All tubes in a tubular hall clock must be stamped "Walter H. Durfee" and must show at least one of his four tube patent dates. The earliest tubes had one patent, dated 1887; the second group had patent dates of 1887 and 1888; the third group, 1887, 1888, and 1892; and the last group, 1887, 1888, 1892, and 1896.

Figure 3.1. Durfee's handstamped name on the top rear edge of the movement. COURTESY OF PAUL FOLEY/DELANEY'S ANTIQUES.

Figure 3.2. Durfee's first 3-weight chiming tall clock, ca. 1887.

- The back of the dial plate and the moondial will be stamped "Walter H. Durfee" at the top of a 1" circle, text at the bottom of the circle is "Providence", and the circle center text is "R.I.".
- Most importantly, the movement must be signed "Walter H. Durfee." In some cases "Walter H. Durfee, Providence" was factory stamped on the lower surface of the backplate by the movement manufacturer during the construction of the movement. This factory-stamp method of identification was primarily used on the 2-weight movements. On the 3-weight movements purchased from J. J. Elliott & Co. of London, Durfee chose not to have the movement stamped "Walter H. Durfee" by the manufacturer.[3.2] Because Durfee sold the tube component package to other clockmakers, he didn't want them to capitalize on his name or reputation. Durfee didn't control the case styles the component buyers used, so he devised a method of identifying clocks by Walter H. Durfee & Co. All movements used in his clocks were signed "Walter H. Durfee" by either the factory or himself. He would handstamp his name on the top rear edge of the backplate (Figure 3.1). All J. J. Elliott movements that Durfee used in his clocks are handstamped "Walter H. Durfee." All authentic Durfee grandfather clocks will be signed in one of these two methods.

Other helpful, but not required, hints to identify a Durfee grandfather clock are as follows:

- "Walter H. Durfee" stamped in a 1" circle on the seat board.
- "Walter H. Durfee" stamped on the nameplate of the dial.
- "Walter H. Durfee" stamped between the two semi-spheres of the dial.

Numbering System

As mentioned in Chapter 2, we will be using Frank Herschede's early 1890 catalog as a basis for numbering the 3-weight tall clock Pattern cases.[3.3] Since we are using Herschede's Pattern numbers, we must point out that Herschede's Pattern numbers are not listed in the chronological order of when they were first produced. It appears that Herschede assigned the numbers randomly, but we can assume that the fancier the case the lower the number.

We will be using Pattern numbers between 2 and 25. Not all 24 Pattern numbers will be used, because research has uncovered only 17 different case styles. Three Patterns will show different variations of one style, such as Pattern 5, Pattern 10, and Pattern 18. When new numbers are used that were not in Herschede's catalog, we have tried to assign numbers in chronological order, based on when we believe new models were first introduced. Our method was based on comparing the chiming-tube dates, any numbers found on the movements, and advertising showing that individual model.

Durfee's First 3-Weight Tall Clock

About 1886, while Durfee was experimenting with his 2-weight tall clock case design, he produced Pattern 35-V1, a variant model (see Figure 2.40, page 28), and a case design for which Durfee had been striving. At the same time, Durfee was working with Harrington and J. J. Elliott & Co. to develop a 3-weight chiming tall clock. We can assume Durfee envisioned that Pattern 35-V1, with a wider 18" waist, would be a perfect case design, because it provided the necessary clearance for Harrington's chiming tubes. According to Elisha Durfee, Walter's nephew and later partner, Walter Durfee chose this basic design for his first 3-weight chiming tall clock (Figure 3.2).[3.4] This clock was Durfee's personal clock and it remained with him and his family for more than 60 years. Elisha Durfee stated, "Wherever Durfee worked or lived he always took his clock with him for it was his pride and joy." This is the only known Durfee 3-weight chiming tall clock with a solid waist door. Durfee quickly envisioned that the public would want to see the pendulum and weights and had beveled waist glasses put in all of his future 3-weight chiming tall clocks.

Here we see that Herschede assigned the Pattern numbers at random, because he called this model Pattern 18, when in reality it should have been Pattern 1. We will cover Pattern 18 in great detail later in this chapter. To make things much simpler, we will list the other 3-weight tall clocks in their numerical and chronological order of production starting with Herschede's Pattern 2. Since there aren't any actual known records of when these clocks were produced, we will attempt to date them within a few years, give or take.

Herschede's Pattern 2 – 1892-1896

Pattern 2 is 104" tall and comes in a beautifully carved solid mahogany case (Figure 3.3). It has wooden carved flaming finials and carved angels above the hood columns (Figure 3.4). The areas above the waist glass (Figure 3.5) and the base (Figure 3.6) have a seashell motif. The case has barley twisted waist columns and carved claw feet.

Herschede's Pattern 4 – 1890-1895

The Pattern 4 solid mahogany case is 104" high, 27" wide, and 17" deep. The case has elaborate carved floral details on the hood, waist, and base (Figure 3.7). The hood has flat fluted columns, fret openings on the sides and special carved-urn finials (Figure 3.8). The waist area has 14 flower rosettes on the front and sides, a carved offset pattern in the trapezoid doorframe, and hand-carved floral detail on the sides in place of columns (Figure 3.9). The base is Bombay shaped with leaf carvings on the front and sides and stands on massive carved claw feet (Figure 3.10). The Harrington chiming tubes have only one patent date, November 8, 1887, mak-

Figure 3.3. Herschede's Pattern 2, mahogany case, ca. 1892-1896. COURTESY OF EDWIN AND LISA BRECHT.

Figure 3.4. Pattern 2, carved hood angels.

Figure 3.5. Pattern 2, carved waist area.

Figure 3.6. Pattern 2, base and claw feet.

ing it one of Durfee's early clocks. The dial is signed "Tiffany & Co. New York". Other features of this Pattern 4 are the door hinges are engraved; the seat board is signed "Walter H. Durfee"; and it has a steel pendulum rod.

Herschede's Pattern 5 – 1888-1905

Pattern 5 is the second most popular 3-weight chiming tall clock model sold by Durfee (Figure 3.11). Normally it stands 100" high, 27" wide, and 17" deep. We have seen a few that were 98" high. Most Pattern 5 clocks had mahogany or oak cases but a few were made in walnut. The hood has a massive carved top with wood finials. There are carved rosettes above the arch dial, it has fret sides, and the hood columns are barley twisted with leaf carvings at the bottom (Figure 3.12). The waist door has 17 beveled lead glass inserts. Some people claim the glass used in these doors came from the Waterford glass factory.[3,5] In some cases the sides have two beveled glass inserts, and in other cases the sides are solid without any inserts. The waist columns are fluted at the top and bottom with figure carvings in the middle (Figure 3.13). The base has a large carved rosette in the center and has claw feet (Figure 3.14).

Herschede's Pattern 5 – V1-1892

We believe that Pattern 5-V1, a variant, was a special order constructed to fit in a room with a ceiling less than 96" tall (Figure 3.15). The clock is basically the standard Pattern 5, except the hood has been made with a fret work top. The casemaker altered the rear of the hood to enclose

Figure 3.7. Herschede's Pattern 4, mahogany case, ca. 1890-1895. COURTESY OF IRA MIZRACH.

Figure 3.8. Pattern 4, hood side view.

Figure 3.9. Pattern 4, waist view.

Figure 3.10. Pattern 4, base and claw feet. COURTESY OF IRA MIZRACH.

Figure 3.11. Herschede's Pattern 5, mahogany case, ca. 1888-1905. COURTESY OF GEORGE TARDIFF (2).

the area where the tube frame is housed. The fretwork top and the tube enclosure can be clearly seen in Figure 3.16. The dial has a "Tiffany & Co., NY" nameplate. The tubes have only one patent date, indicating that the clock was made in the early 1890s.

Herschede's Pattern 8 – 1890-895

The only known Pattern 8 was sold in a New England auction in 1982 to an owner who wanted to remain anonymous. We learned only that the movement was signed "Durfee". Figure 3.17 was taken from Herschede's catalog,[3.6] while Figure 3.18 shows the auction house image. Note that the base in Figure 3.18 has three paneled inserts. The auction house stated that it had "Walter H. Durfee, Providence, RI" on the dial and all tubes were marked "Durfee". Note that the door has 17 beveled glass inserts, the same number as Pattern 5.

Herschede's Patterns 3, 6, and 7

These three Patterns have never been seen either in private collections or at auction houses. Until they are seen we will not know whether they were produced by Herschede or Durfee. Therefore, we have removed these Patterns from the Durfee list of 3-weight tall clocks.

Durfee's First 3-Weight Hall Clock Advertisement – 1888

In late 1888 Durfee published his first known advertisement notifying the public that he was a manufacturer and importer of English hall clocks (Figure 3.19). The ad

Figure 3.12. Pattern 5, close-up of hood.

Figure 3.13. Pattern 5, waist.

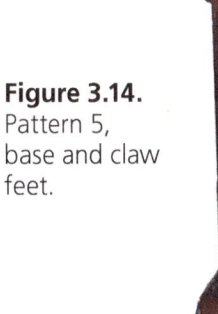

Figure 3.14. Pattern 5, base and claw feet.

Figure 3.15. Herschede's Pattern 5-V1, variant, ca. 1895. COURTESY OF DENNIS JACOB.

Figure 3.16. Pattern 5-V1, hood.

states that he had complete control of the US rights of the Harrington patent tubular bells, and it also listed the eight agencies selling his clocks. The ad included a frontal and rear view of a 3-weight chiming hall clock not included in Herschede's catalog.

Pattern 9 – 1888-1895

We assigned the Pattern 9 designation to the clock in Durfee's 1888 advertisement. The actual clock (Figure 3.20) stands 104" high and came in a beautifully carved oak case. The case has an unusual hood and top (Figure 3.21). The top of the hood is supported by 21 double-tapered 5" high spindles on the front and 16 spindles on each side. Above the spindles is a heavily carved fan in the shape of raised-arched sunburst, which is surrounded by two cathedral-steeple finials (Figure 3.22). A close-up of the seashell carvings, relief sunburst, and the gargoyle can be seen in Figure 3.23. The waist columns have twist carvings on the lower portion and tulip-shaped capitals. The base is trapezoid shaped with a trapezoid cutout in the center of the base (Figure 3.24). The feet are not the standard claw feet but are simply a flat surface with flower-like carvings. On the inside of the case just below the seat board, written in pencil, is "Sold by Holland & Derby, January 1, 1892 to Henry E. Conant, Concord, N.H." Holland & Derby is listed in the 1890 *New Hampshire Directory* as selling watches and jewelry at 54 N. Main St., Concord, NH.

Our research has uncovered only four of Pattern 9 and its variant Pattern 9-V1-style cases. Three of them have the J. J. Elliott & Co. movement with Westminster chime on four flat-wire gongs and the Whittington chime on a nest of eight bells. The others have a standard J. J. Elliott 3-weight chiming movement.[3.7] The J. J. Elliott factory stamped these movements for Durfee. The oval-shaped

Figure 3.17. Herschede's Pattern 8, ca. 1890-1895. This image is Figure 351 on page 151 of Tran Duy Ly's book titled *Longcase Clocks and Standing Regulators* from arlingtonbooks.com.
COURTESY OF ARLINGTON BOOKS.

Figure 3.18. Pattern 8 auction house photo 1982.

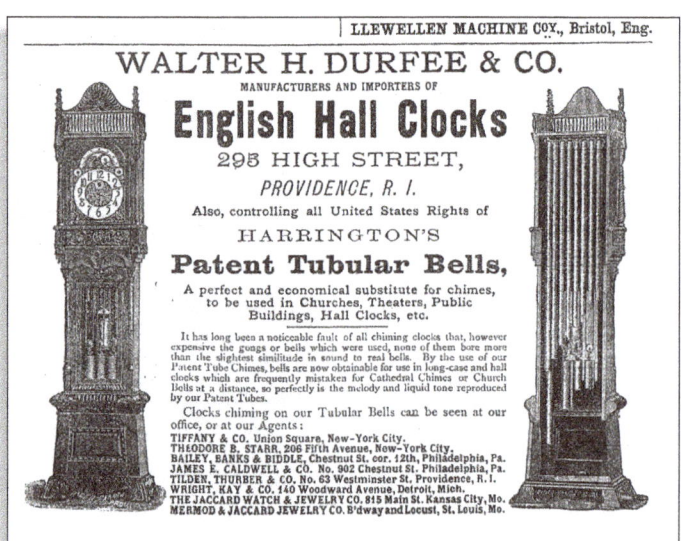

Figure 3.19. Pattern 9 shown in Durfee's 1888 ad.

Figure 3.20. Pattern 9 tall case clock, ca. 1888-1895.
COURTESY OF ANDREW APICELLA.

Figure 3.22. Pattern 9, close-up of top.

Figure 3.23. Pattern 9, area above waist door.

Figure 3.25. Pattern 9-V1 tallcase clock.

Figure 3.21. Pattern 9, hood and top.

Figure 3.24. Pattern 9, base and feet.

Figure 3.26. Pattern 9-V1, hood and top.

Walter H. Durfee—Chapter 3 • 45

Figure 3.27. Pattern 10 ad and letterhead sketch. HARPER'S MAGAZINE ADVERTISER, CA. 1890-1896.

factory stamp on the back of the movement is 5/8" long and 7/8" wide and reads "Waltr H. Durfee & Co" at the top and "Providence, RI" at the bottom. The name "Walter" was spelled "Waltr". The factory also stamped an identification number on each movement; the lowest number was 126 in the Pattern 9 clock (Figure 3.20) and the highest number was 135, which was found in a Pattern 9-V1 variant.

Only one Pattern 9 case with the tubular bell chime movement, shown in Durfee's 1888 advertisement (Figure 3.19), has ever been found. All four Pattern 9 cases mentioned in this section have dial nameplates that read "Walter H. Durfee, Providence, RI."

Pattern 9-V1 – 1890-1895

One of the Pattern 9 tall clocks mentioned is variant Pattern 9-V1 with a solid mahogany case that must have been a special order (Figure 3.25). The height of the clock was altered to 96" so it could fit in a room with a low ceiling. Durfee had the casemaker remove the 53 5" spindles and the support braces, which shortened the case height by 8" (Figure 3.26). The two cathedral steeples were altered as well.

Pattern 10 – 1890-1896

Figure 3.31. Pattern 11, base and claw feet.

Figure 3.29. Pattern 11, hood top and angel columns. ANONYMOUS.

Figure 3.30. Pattern 11, area above waist door.

Figure 3.28. Pattern 11 tallcase clock, ca. 1890-1900. COURTESY OF MAURICE BURNS (4).

Pattern 10 was shown in several Durfee advertisments and on some of the letterhead of his business stationery (Figure 3.27). We have never seen nor heard of an actual clock, but we think that it should be assigned a Pattern number.

Pattern 11 – 1890-1900

Pattern 11 (Figure 3.28) came in a solid mahogany case and stands 100" high. The case is heavily carved from top to bottom. The top of the hood has a removable crest carving and portrays the North Wind, surrounded by a flower-pattern motif. The hood's two side columns are two full-length angels with their wings extended onto the face and sides (Figure 3.29). The case has a recessed carving below the hood and just above the waist door (Figure 3.30). This carving is similar to the waist area in Patterns 20 and 41. The fluted waist columns are flat and have carved capitals and bases. The top, center, and lower part of the base are heavily carved, with the base having large claw feet (Figure 3.31). The elaborate carved case came with a one-jar mercury pendulum. The nameplate is missing but the movement, tubes, and back of the dial are signed "Walter H. Durfee".

Figure 3.33. Pattern 12, hood and top.

Figure 3.34A. Pattern 12, carved waist columns capital.

Figure 3.34B. Pattern 12, heavily carved base and claw feet.

Figure 3.32. Pattern 12 tallcase clock, ca. 1891-1900. COURTESY OF LYNDHURST/NATIONAL TRUST.

Pattern 12 – 1891-1900

Pattern 12 (Figure 3.32) has an oak case that stands at 102". The hood has trapezoid columns with a massive large carving at the top crest (Figure 3.33). The waist columns have two carved ladies' heads at the top (Figure 3.34A). The base is heavily carved from top to bottom, including the claw feet (Figure 3.34B). This Pattern 12 case came with a three-jar mercury pendulum.

Pattern 13 – 1892-1894

Stationery letterhead used by Walter H. Durfee & Co., after 1900, had a sketch of a Durfee tubular chime clock, which never surfaced. Therefore, for many years, we believed the clock was never made or produced. In 2012 the clock surfaced and after close examination we found that it was a true Durfee 3-weight tall clock and labeled it Pattern 13 (Figure 3.35A). The solid mahogany case stands 100" high and is 26-1/2" wide and is beautifully carved with marquetry inlay on the hood and base. The tubes have only three patent dates indicating that it was made sometime around 1892-1894. The dial is one of a kind with scalloped pierced inserts in place of common square corners (Figure 3.35B). The base has Chippendale-style feet in place of claw feet (Figure 3.35C).

Figure 3.35A. Pattern 13 tallcase clock, ca. 1892-1894. COURTESY OF ERIC AND PAULA LITSCHER.

Figure 3.35B. Pattern 13, unusual dial.

Figure 3.35C. Pattern 13, base.

Figure 3.36. Pattern 14 tallcase clock, ca. 1892-1900. COURTESY OF JEROME AND SUSAN CIULLO.

Figure 3.37. Pattern 14, hood and top.

Figure 3.38. Pattern 14, waist and 2-jar mercury pendulum.

Pattern 14 – 1892-1900

The Pattern 14 case shows a major change in Durfee's case designs (Figure 3.36). Many previous Durfee's Pattern cases were more massive in size and carvings. Some clocks overpowered a standard room. Starting with Pattern 14, Durfee's case designs had a more American tall clock case style, resulting in a softer more appealing appearance. Pattern 14 stands 101" high with a solid mahogany case that has glass sides. The hood has unfluted solid columns, wood finials, and much sharper and more detailed carvings than what we had seen on prior Patterns (Figure 3.37). The waist door is oval shaped, while the waist area surrounding it has softer and more detailed carvings (Figure 3.38). The waist has wooden columns, wooden capitals, and beveled glass sides. The base has a reverse-Bombay shape with a delicate floral design, while the feet are carved angel faces with spread wings (Figure 3.39). This Pattern 14 has a two-jar mercury pendulum.

Pattern 15 – 1892-1900

In Pattern 15 Durfee's case design again drifted toward softer lines (Figure 3.40). Pattern 15 has a solid mahogany case that is 98" high, 27" wide, and 17-1/2" deep. The hood has wooden finials, added leaf carvings to the swan necks, and half fluted columns with wooden capitals and bases (Figure 3.41). The waist has an oval top door, simple flower carvings above the door, and quarter-round fluted columns with wooden capitals and bases. The base has much simpler lines with only a carving of crossed torches in the center section of the base (Figure 3.42). The dial nameplate reads "Bailey, Banks & Biddle" and "Philadelphia". The "Walter H. Durfee" stamped movement, the back of the dial, and the pendulum have the number "136" stamped on them. The brass pendulum bob has the initials "B.K.Jr and L.G.K." and the date "1898". We have seen a few other Durfee pendulum bobs with either a name or initials and dates; in

Figure 3.39. Pattern 14, base and angel face feet.

Figure 3.41. Pattern 15, hood and flaming wooden finials.

Figure 3.40. Pattern 15 tallcase clock, ca. 1892-1900. COURTESY OF ANDREW APICELLA.

Figure 3.42. Pattern 15, base and claw feet.

Figure 3.43. Pattern 16 102" tallcase clock, ca. 1892-1900. COURTESY OF JEROME AND SUSAN CIULLO (4).

Figure 3.44. Pattern 16, hood and acorn-shaped center finial.

Figure 3.45. Pattern 16, waist.

Figure 3.46. Pattern 16, base and ram's horns and hooves.

most cases, this clock was given as a wedding gift. The tubes have only three Durfee patent dates, so we can assume the clock was made between 1892 and 1896; 1898 is more likely the date the clock was sold.

Pattern 16 – 1892-1900

Pattern 16 is 101" high, 30" wide, and 20" deep, and came in a solid mahogany case, which has beveled glass sides (Figure 3.43). The Pattern 16 hood has an acorn center finial with carvings below the swan's neck pediment, and it has half-fluted solid columns with wooden capital and bases, similar to the columns in Figure 3.44. The waist area has fine detailed carvings above and below the door, while the side columns are a larger version of the hood columns (Figure 3.45). The base has a beautiful carved center panel offset by a relief center frame panel, while the feet are in the form of ram's horns and hooves (Figure 3.46). The dial nameplate is signed "Tiffany & Co., New York".

Pattern 16 also came in oak cases with solid columns on the hood and waist. In some cases the center acorn-design finial was replaced with three wooden finials

Figure 3.47. Pattern 17 tallcase clock, oak case, ca. 1890-1892. COURTESY OF SOUTH HAMPTON ANTIQUES.

Figure 3.48. Pattern 18 tallcase clock, mahogany case, ca. 1887-1912. COURTESY OF PAUL FOLEY/DELANEY'S ANTIQUES.

Figure 3.49. Pattern 18, hood and urn-shaped wooden finials. COURTESY OF PAUL FOLEY/DELANEY'S ANTIQUES.

Figure 3.50. Pattern 18, base and claw feet.

across the top. After 1915 Waltham Clock Co. used this case design, with Waltham tubes and movements, for a few years.[3.8]

Pattern 17 – 1892-1900

The true Durfee Pattern 17 stands 101" high and came in an oak case with wooden urn finials and claw feet (Figure 3.47). The hood columns consist of a cluster of three pillars with wooden capitals and bases. The waist has fluted columns with carved capitals and bases. The dial nameplate is "Jas. E. Caldwell & Co., Philadelphia". The tubes only have three patent dates, which indicates the clock was made before 1897.

Herschede's Pattern 18 – 1887-1912

The Herschede catalog shows this clock as Pattern 18, although Durfee may have had a different number for it. Because it is normally referred to as Pattern 18, we will do the same.

Pattern 18 was Durfee's best seller and it appears that two-thirds of all 3-weight chiming tall clocks he produced were the Pattern 18 model or one of its variants.

Although all Pattern 18 cases are the same, they varied in the type of columns, capitals, and finials used. Besides the original Pattern 18 there were three distinct variations, and we will discuss each one individually.

Pattern 18 – 1887-1912

The basic design of Durfee's Pattern 18 (See Figure 3.2 on page 40) that he first made in 1887 remained the same for 25 years.

Figure 3.48 is a standard Pattern 18 that has six full wooden columns. Figure 3.49 shows the hood wooden columns, the wooden capitals and bases, and wooden flame finials. The base has inset front and side panels plus the normal claw feet (Figure 3.50). The solid wood cases came in light mahogany color, normal mahogany, dark mahogany color, oak, and walnut. The cases had raised waist and base panels, carved swan's neck pediments, double-cut beveled glass panels, ornamental lions-head escutcheons, brass hinges (some were engraved), a half-circle fan carving at the top of the waist door, and all movements were stamped "Walter H. Durfee" on the top edge of the rear plate. Sometimes the top edge was stamped twice, once on the right side and in reverse or the opposite direction on the left side. The tubes would have one, two, three, or four patent dates depending on when the clock was made. All quarter-hour chimes struck either the Westminster or eight bell tunes.

Pattern 18-V1 – 1890-1912

Pattern 18-V1 (Figure 3.51) is identical to Pattern 18, except it has six fluted columns with brass capitals and bases, brass finials, and two swan's neck brass trim cover plates. Figure 3.51A shows the fluted columns that have brass capitals and brass bases, brass finials, and brass swan's neck trim.

Pattern 18-V2 – 1892-1910

Pattern 18-V2 is identical to Pattern 18-V1, except the six columns are barley twisted rather than fluted (Figure

Figure 3.51. Pattern 18-V1 tallcase clock. COURTESY OF JEROME AND SUSAN CIULLO.

Figure 3.52. Pattern 18-V2, barley twist columns, ca. 1892-1910. ANONYMOUS.

Figure 3.51A. Pattern 18-V1, hood with fluted columns, brass capitals, and bases.

Figure 3.53. Pattern 18-V2, hood.

3.52). Figure 3.53 shows the hood's barley twist columns and brass finials. The waist-door fan, twisted columns, and brass capitals and bases can be seen in Figure 3.54.

Pattern 18-V3 – 1900-1912

Pattern 18-V3, the third variation of the original Pattern 18, wasn't marketed by Durfee until the turn of the twentieth century (Figure 3.55). The four major variations from Pattern 18 consist of the following:
- The waist door opening was made longer by omitting the fan-shaped carving at the top. This allowed for a larger beveled glass opening.
- At the top of the glass opening fretwork was added, giving the illusion that the waist glass had ten glass inserts (Figure 3.56).
- The hood silk sound-cloth openings were replaced with glass panels (Figure 3.57).
- The prior habit of only using brass finials with brass capitals and bases was disregarded, and Durfee chose to use wooden flame finials in their place.

Pattern 19 – 1895-1900

Pattern 19 stands 101" high and came in a solid mahogany case (Figure 3.58). The fan carving was replaced with a floral pattern (Figure 3.59). The case has six barley twisted columns with carved capitals and bases. The base has nine inset panels (Figure 3.60).

Figure 3.54. Pattern 18-V2, waist area. ANONYMOUS.

Figure 3.56. Pattern 18-V3, waist door fretwork.

Figure 3.57. Pattern 18-V3, side view of hood and top.

Figure 3.55. Pattern 18-V3 tallcase clock, ca. 1900-1910. COURTESY OF GEORGE TARDIFF (3).

Figure 3.58. Pattern 19 tallcase clock, ca. 1895-1900. COURTESY OF JAMES LAPINSKY. (3)

Figure 3.59. Pattern 19, top carving on waist door.

Figure 3.60. Pattern 19, base with nine inset panels.

Figure 3.61. Pattern 20 104" tallcase clock, ca. 1888-1910. COURTESY OF GEORGE TARDIFF.

Figure 3.62. Pattern 20, hood and top with North Wind carving.

Herschede's Pattern 20 – 1888-1910

We believe that Pattern 20 was one of Durfee's earlier models and was first produced around 1888, when he was using massive and heavily carved cases, such as Pattern 2 through Pattern 8. This Pattern 20 stands 104" high and came in a solid mahogany case (Figure 3.61). Oak Pattern 20 cases outnumber mahogany cases two to one. The top of the hood has a removable, detailed, figure carving of Satan (Figure 3.62). The waist area is heavily carved, and the door glass opening comprises 13 beveled lead crystal inserts (Figure 3.63). Figure 3.64 shows the detailed carvings in the base plus the claw feet.

Pattern 21 – 1898-1905

Pattern 21 appears to be one of a kind because we have only seen the one shown in Figure 3.65. The mahogany case doesn't have many of the features we have seen in the other Patterns, but the clock is all original, and has a Durfee-stamped movement, tubes, dial, and nameplate. The case is nicely detailed, while the hood has eight unusual columns, side panels, swan's neck pediment, and center finial (Figure 3.66).

Patterns 22, 23, & 24

We deliberately left these Pattern numbers open, so they could be assigned to Durfee 3-weight clocks as they are discovered.

Pattern 25 – 1905-1910

It is difficult to determine when Pattern 25 was made, because it doesn't fit into any common Durfee design. It appears that it was made after 1902, the year Durfee lost his chiming-tube patent rights, was facing a competitive pricing market, and needed to reduce the cost of his cases (Figure 3.67). The oak case stands 100" high and has simple hood lines, but the case is heavily carved on the hood sides, waist, and base (Figure 3.68). The base is a simple design with a carved insert and Chippendale feet (Figure 3.69).

Figure 3.63. Pattern 20, waist door with 13 beveled glass inserts.

Figure 3.64. Pattern 20, base and claw feet. COURTESY OF GEORGE TARDIFF.

Figure 3.65. Pattern 21 tallcase clock, ca. 1898-1905. COURTESY OF SUNDIAL FARM ANTIQUES (2)

Figure 3.67. Pattern 25 tallcase clock, heavily carved oak case, ca. 1905-1910. ANONYMOUS (3).

Figure 3.68. Pattern 25, full view from the side.

Figure 3.66. Pattern 21, hood and top center ball finial.

Figure 3.69. Pattern 25, carved base insert and Chippendale feet.

Walter H. Durfee—Chapter 3 • **55**

Three-Weight Tall Clock Production Comparisons – 1887-1910

Walter H. Durfee & Co. only manufactured approximately 400 tall clocks. Of those, only about 275 were the 3-weight chiming hall clocks produced from 1888 to 1910. Because no known records were kept of actual clocks sold, we will attempt to determine the number of models made and their rarity. The following figures are estimates based on the number of each model of which we have records, comparing individual numbers to the total recorded. These numbers will be averaged and rounded off with a margin of error of 5 to 10 percent and are as follows:

- Pattern 18 and its variants account for the largest volume produced at approximately 66 percent, or 180 clocks.
- Pattern 5 was the second most popular model and it accounts for approximately 10 percent, or 30 clocks.
- Pattern 20 accounted for approximately 5 percent, or 15 clocks.
- The remaining 13 Patterns accounted for 50 clocks, or approximately 4 or 5 per Pattern.

Walter H. Durfee & His Clocks

Chapter 4

Three-Weight Tall Clocks' Movements, Weights, Dials, Cases, and More

This chapter continues our research on Walter H. Durfee & Co.'s 2- and 3-weight tall clocks. Here we cover in detail the 3-weight tall clock movements, dials, tubes, pendulums, weights, finials, feet, hardware, sales agents, and advertisements.

Movements

Our research has shown that Durfee only used J. J. Elliott & Co., London, England, movements in his 3-weight tall clocks. However, some people have said they had or had seen a Durfee 3-weight tall clock with a German Winterhaler & Hofmier, Elite, or other movement. When tracking these statements to the actual clock or to its owner, we found in all cases that these clocks were not made by Durfee. Dana Blackwell,[4.1] a renowned US authority on tall clocks with 60 years of cleaning and servicing tall clock movements, said Durfee only used J. J. Elliott & Co. movements in all of his 3-weight tall clocks. Henry Fried,[4.2] while researching his April 1982 *NAWCC Bulletin* article on tall clock chime tunes, said he had never run across anything except a J. J. Elliott & Co. movement in any of Durfee's 3-weight tall clocks. Most importantly, Elisha Durfee, Walter's nephew and helper, verified that Durfee only used J. J. Elliott & Co. movements in his 3-weight tall clocks. Per Tom Splitter, an Elliott & Co. authority, Durfee used not only their nine tube chiming movements but also their 4-gong and eight bell chime movement as well as their 4-gong chime movement.[4.3] The illustrations of these three movements in Tran Du Ly's book titled *Long Case Clocks and Standing Regulators Part I: Machine Made Clocks* were taken from Herschede Clock Co.'s 1985 trade catalog. In the catalog the images of the movements are inverted. Figures 4.1, 4.6, and 4.9 in this chapter present the images correctly, correcting a more than a century-old error.

Tubular Bell Chime Movement

Used in at least 90 percent of all Durfee 3-weight tall clocks, the tubular

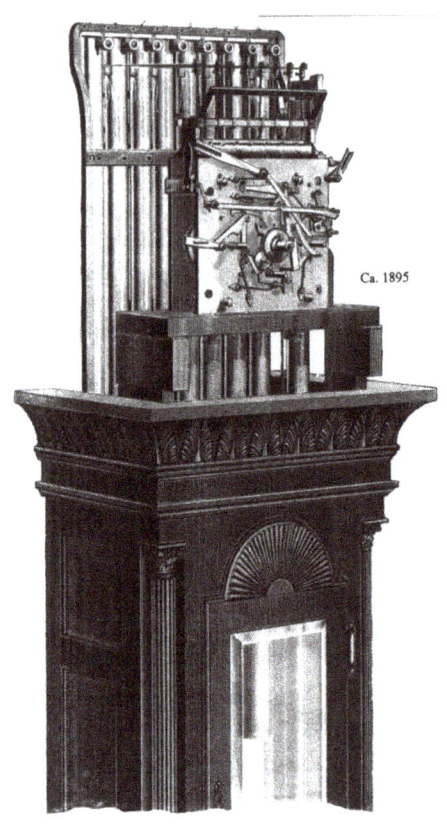

Figure 4.1. J. J. Elliott's tubular bell chime movement. This image is Figure 338 on page 144 of Tran Duy Ly's *Longcase Clocks and Standing Regulators* from arlingtonbooks.com. COURTESY OF ARLINGTON BOOKS.

bell chime movement was the most popular movement (Figure 4.1). The J. J. Elliott movement is capable of chiming on the Harrington tubes[4.4] at each quarter hour, playing either the Westminster, Whittington, or 8-bell tunes. The hour is struck on a separate tube. Most J. J. Elliott movements (Figure 4.1) have the 8-bell chime. It wasn't until 1893-1894 that Durfee offered the Whittington chimes and the 8-bell chimes. The nickel-plated tubes are suspended on a crossbar that ensures the proper position with respect to the hammers that strike the tubes, thus retaining an even quality of tone. The movement is constructed so that by turning a small pointer on the dial, the chime is made silent and the hour tube only is struck, or both can be shut off.[4.5] The front view of the movement shows the location of the chime cylinder and the rack holding the 8-bell chime hammers (Figure 4.2). Figures 4.3 and 4.4 show the movement mounted to the dial, the location of the hour strike hammer, and the two fans mounted to the outside of the rear plate. These J. J. Elliott movements are always signed "Walter H. Durfee" on the top edge of the rear plate (Figure 4.5). In some cases Durfee handstamped his name twice on the top edge of the rear plate.

Four-Gong and 8-Bell Chime Movement

The 4-gong and 8-bell chime movement (Figure 4.6) was used in Durfee's Pattern 9 (see Figure 3.20 in Chapter 3). Depending on the year the clock was made, the 8-bell chimes or the Whittington chimes are rendered upon the eight brass nickel-plated bells at the top of the movement (Figure 4.7). The Westminster chimes on the four wire gongs is located behind the movement, while the hour is struck on a wire gong just behind the Westminster gongs (Figure 4.8). The gongs are made of drawn flat steel wire in a spiral shape and are mounted to an improved sounding box, designed by J. J. Elliott, which gives them a full and free vibration and renders the tone of even quality. The variation in the tenor of the two melodies makes it desirable. The andante, a moderately slow tempo, of the four wire gongs Westminster chime has a solemn and imposing nature, while the vivace, a lively and fast tempo, of the eight jin-

Figure 4.2. Front view of movement shown in Figure 4.1. COURTESY OF PAUL FOLEY/DELANEY'S ANTIQUES.

Figure 4.3. Rear view of chime side of movement shown in Figure 4.1. COURTESY OF PAUL FOLEY/DELANEY'S ANTIQUES

Figure 4.4. Rear view of strike side of movement shown in Figure 4.1.

Figure 4.5. "Walter H. Durfee" ID stamp on top edge of rear plate.
COURTESY OF PAUL FOLEY/DELANEY'S ANTIQUES.

Figure 4.7. Chime side view of movement shown in Figure 4.6.
COURTESY OF JACK DAVIS (2).

Figure 4-9. J. J. Elliott 4-gong chime movement. This image is Figure 340 on page 145 of Tran Duly Ly's *Longcase Clocks and Standing Regulators* from arlingtonbooks.com. COURTESY OF ARLINGTON BOOKS.

Figure 4.6. J. J. Elliott 4-gong and 8-bell chime movement. This image is Figure 339 on page 145 of Tran Duly Ly's *Longcase Clocks and Standing Regulators* from arlingtonbooks.com.
COURTESY OF ARLINGTON BOOKS.

Figure 4.8. Strike side view of movement shown in Figure 4.6.

gling Whittington bells inspires the soul to the pleasures of life.[4.6] The chime or bell action as well as the hour strike can be made silent by turning levers on the dial.

Four-Gong Chime Movements

The 4-gong chime movement was only used in Durfee's latter stages of production (Figure 4.9). From the front view it appears to be a typical 8-day time-and-strike movement (Figure 4.10). This movement is basically the 4-gong and 8-bell chime movement, except the 8-bell chime mechanism has been omitted. The Westminster chime is rendered on four gongs at every quarter hour.

The hour is struck on a large wire gong stationed in back of the chime flat wire gongs, which are both fastened to a sounding box. The chime strike, the hour strike, or both can be put in action or silenced at will.[4.7] The rear view of the movement shows the Westminster chime cylinder, the four quarter-hour strike hammers, and the hour gong hammer (Figure 4.11).

Movement Identification Stamps

J. J. Elliott & Co. didn't put any factory identification marks or stamps on the tubular bell chime movements Walter H. Durfee & Co. purchased, apparently as part of

Figure 4.10. Front view of Figure 4.9 movement. COURTESY OF RICHARD OLIVER/JOHN TANNER.

Figure 4.11. Rear view of movement in Figure 4.9. COURTESY OF RICHARD OLIVER/JOHN TANNER.

Figure 4.12. Oval ID stamp on Figure 4.6 movement.

Figure 4.13. Circle ID stamp on Figure 4.9 movement.

their agreement. This arrangement allowed Durfee to take full credit for making the movements. Durfee even referred and boasted to his clientele, associates, family, and friends about "his" movement factory in England. However, he never owned any actual manufacturing facilities in England. Durfee purchased all his 2- and 3-weight movements from established movement makers in England. In the tubular bell chime movements, Durfee stamped his name on the top edge of the rear plate when the movement reached his shop in Providence, RI (see Figure 4.5).

The aforementioned agreement between Elliott and Durfee did not apply to the Elliott 4-gong and 8-bell movements. As was pointed out in Chapter 3, this movement was only used in Durfee's Pattern 9 clock case, one of the first 3-weight chiming tall clocks Durfee produced as early as late 1887 or early 1888. J. J. Elliott & Co. factory stamped these movements for Durfee. The oval factory stamp on the back of the movement is 5/8" long and 7/8" wide and reads "Waltr H. Durfee & Co."—note the spelling of the first name—at the top and "Providence, RI" at the bottom (Figure 4.12). Our research has located only four Pattern 9 chiming tall clocks, and of the four, three used the Elliott 4-gong and 8-bell movements, which have the previously described Durfee factory stamp. The factory also stamped an identification number on each movement; the lowest number was 126 and the highest, 135. It appears that Durfee stopped ordering these movements as early as 1883-1884 and concentrated solely on producing tubular clocks for the next 18-20 years.

The 4-gong chime movement (Figure 4.9) was found in only one Durfee clock, and it appeared to have been manufactured after 1905. The rear edge of the movement is stamped "Walter H. Durfee" and a 1" round dial stamp is on the lower corner of the backplate (Figure 4.13). Whether Durfee was downsizing his operation or trying to compete with a lower quality market, we may never know.

Dials in General

The identities of the actual English dial manufacturers or suppliers Durfee used still remain a complete mystery to us. Our research did not uncover any insights or leads on English dial makers or dial suppliers during this period of time.

The exact method Durfee used in selecting a 13" brass dial for his 3-weight tall clocks is not known for sure either. However, we do know that Durfee only used the hand-pierced dial[4.8] and the hand-engraved dial.[4.9] Each dial style had different grades with some grades costing three and four times more than the lowest grade. It appears that Durfee used the following methods to obtain the dials:

- He could have purchased the movement from J. J. Elliott with an attached dial; the actual dial style and grade would have been selected either by Elliott or by Durfee.
- He could have purchased the dials and supplied them to J. J. Elliott, who attached them to the movement.
- He could have purchased the movement without a dial and added the dial later.

We may never know which method Durfee used, but we do know he made sure the two different style dials were interchangeable with each other. This interchangeability gave Durfee the flexibility of replacing dials in his Providence, RI, shop and maintaining an inventory of five or six different dials at one time.

Pierced Dials

The most common dial Durfee used was the pierced dial, which also was the most expensive dial produced (Figure 4.14). The dial brass plate is extra heavy and is highly polished and lacquered. The decorative four corners, center ornament, and the small insert between the hemispheres are hand pierced (sawed) and hand engraved and are richly gold plated. The raised hour numerals are highly finished and gold plated. The raised and engraved hour and minute circles, the two chime circles, the two hemispheres, and lunar arch are made of heavy brass and are richly silver plated. The moondial was always hand painted. The nameplate above the 6 was silver plated and contained the seller's name, in this case "Walter H. Durfee & Co, Providence."

These hand-pierced dials could be purchased with different features. In some cases the center pierced insert was omitted and the center section of the dial plate was hand engraved. In other cases, the raised numerals were beveled or the hemisphere hand engravings were revised from a globular to a landscaping or nautical scene.

The most expensive and beautiful Durfee dial that we have found has pierced, engraved, and embossed four corners, plus an embossed center insert (Figure 4.15). In addition, the two engraved global spheres are finely detailed; the raised engraved chapter ring has gold-plated numerals, with each number surrounded by an embossed gold engraved ring. The strangest Durfee dial must have been a special order, because we have never seen these unusual designed gold-plated numbers on any other Durfee clock (Figure 4.16).

Depending on the degree of quality and amount of hand engraving, the cost of a hand-pierced dial could increase the initial price of a $500 clock by $150-$200, or more.

Engraved Dials

The hand-engraved brass dials were less expensive than the pierced dials because the latter did not require the tedious labor involved in hand piercing the corners and center insert. The extra heavy dial plate surface was beautifully engraved in the four corners and the center section (Figure 4.17). The hour and minute circles, the two chime rings, the two hemispheres, and lunar arch were all raised, engraved, and silver plated. The raised numbers are gold plated. The degree and quality of the engravings on the dial would determine its price.

Moondial

The moon wheel has 118 teeth and moves one tooth every 12 hours; therefore, in 59 days it accomplishes a full revolution of the wheel or two full-moon rotations. The figures in the lunar arch go up to 29-1/2, which represents the length of a lunar month or one full-moon rotation.

The moondials Durfee used had two hand-painted moons that had large detailed lips, a pug nose, heavy eyebrows, big eyes, and rosy cheeks. In the early years the moon had a pointer on the top of the moon, making it easier to read the moon calendar (Figure 4.18).

The moondial wheel is mounted on a post on the back of the dial plate and can be easily removed for servicing or cleaning (Figure 4.19). The moondial wheel contains two full moons and two scenes opposite of each other. The two painted scenes are mostly a landscape scene on one side and a nautical on the other side (Figure 4.20). In the early years of the 3-weight moondials, the two silver-plated hemisphere inserts were mostly hand-engraved

Figure 4.14. Example of a Durfee hand-pierced dial.

Figure 4.15. Example of a Durfee pierced and embossed dial.

Figure 4.16. Unusual dial numerals.

Figure 4.17. Durfee engraved dial.

Figure 4.18. Durfee moondial.

Figure 4.19. Rear view of a Durfee moon wheel.

Figure 4.20. Front view of a Durfee moon wheel.
COURTESY OF PAUL FOLEY/DELANEY'S ANTIQUES.

globes (Figure 4.21). In later years the hemisphere globes were replaced with engraved landscape and nautical scenes (Figure 4.22). The back of the moondial and the dial plate were always stamped "Walter H. Durfee Providence RI" in a 1" circle (Figure 4.23).

We assume the face of the moondial was hand painted after the Durfee stamp was applied. Putting Durfee's 1" circle stamp on after it had been painted could cause the hand-painted surface to chip or flake on the opposite side of the stamp. This brings up the questions: Who put the stamp on the dial? Was it the dial maker, the dial supplier, the movement maker, or Durfee himself? If it was the dial maker or supplier, they would have needed to have the Durfee stamp to stamp the dial prior to having the moondial painted. Durfee could have used different dial makers, and all would have had to have a stamp. It appears that the movement maker never got involved. Our theory is that Durfee purchased and received the moondials unpainted; he handstamped them himself; and Durfee then had the moondials hand painted by someone in the Rhode Island or the New England area.

Dial Nameplates

Durfee dials had a nameplate clearly showing the seller or jewelry store agent who sold the clock. A few examples are as follows:

- "Walter H. Durfee, Providence. R.I.", who more than likely sold the clock directly to a customer (Figure 4.24)
- "Bailey, Banks & Biddle Co., Philadelphia" (Figure 4.25)
- "Theodore B. Starr, New York" (Figure 4.26)
- "Tiffany & Co" (Figure 4.27)
- "Tiffany & Co. New York" (Figure 4.28).

By a large margin Tiffany & Co., New York, sold the largest number of Durfee's 3-weight tall clocks. Tiffany and Durfee must have had a special agreement, because Tiffany & Co. was only listed as a Durfee agent in 1888.

Figure 4.21. Engraved global hemispheres. COURTESY OF JEROME AND SUSAN CIULLO.

Figure 4.24. "Walter H. Durfee" stamped on nameplate. COURTESY OF JEROME AND SUSAN CIULLO.

Figure 4.22. Engraved nautical and landscaping hemispheres. COURTESY OF JEROME AND SUSAN CIULLO.

Figure 4.25. "Bailey Banks & Biddle" stamped on nameplate. COURTESY OF ANDREW APICELLA.

After 1888 Durfee never listed Tiffany & Co. as one of his sales agents, but it sold several Durfee 3-weight tall clocks from 1888 through 1902. One of the rarest dial identification markings is the engraved name of the seller, "Walter H. Durfee Providence" in the embossed area of the center section of the dial (Figure 4.29).

Harrington Chime Tubes

In 1887 Durfee received the sole rights to John Harrington's US Patent No. 372,849, dated November 8, 1887, creating an apparent chiming tube clock monopoly for Durfee. The Harrington patent was for a "Chiming Apparatus for Clocks" not a "Tube" manufacturing patent. Durfee claimed that he not only had rights to manufacture Harrington's tubes but he also had the rights to control the uses of tubular bells in any clock manufactured in the United States. This meant that if any business or person wanted to manufacture a chiming tubular clock, they had to get his permission. To secure his permission, Durfee required that they purchase the tubes directly from him. In many cases he insisted that they purchase other clock components such as the movements, dials, and weights.[4.10]

The Harrington patent rights did not require Durfee to purchase his tubu-

Figure 4.23. Durfee's ID stamp on back of moon wheel with photo inset of detail.

lar chime tubes from Harrington. The agreement allowed Durfee to purchase tubes from other sources, as long as they were located in the United States. It appears that Durfee had made prior arrangements with US Tubular Bell Co. in Methuen, MA,[4.11] to be his American Harrington tube supplier. By the time Harrington received his official US patent, Durfee and US Tubular Bell Co. had tubes available. The only work left for US Tubular Bell Co. was to stamp the patent date, "Nov. 8, '87", on the tubes and ship them to Durfee in Providence, RI.

For the next 10-12 years Durfee dominated the 3-weight hall clock market in the United States. Not everyone was happy with the Harrington and Durfee agreement. As early as 1890 Durfee had many disgruntled clock manufacturers fighting him and his Harrington patent rights.

From 1887 to 1896 Durfee obtained three more patent rights on the Harrington tubes. The new patent rights never really changed or altered the tubes US Tubular Bell Co. manufactured. Durfee used the patent dates as "cosmetic dressing" on the tubes to impress and to discourage other clock manufacturers from questioning his patent rights. Durfee's chime tubes are all marked "Walter H. Durfee & Co. Prov. R.I." and will have either one, two, three, or four patent dates stamped on them. The four dates are "Nov. 8, '87"; "Sep. 18, '88"; "Nov. 1, '92"; and "Oct. 6, '96" (Figure 4.30). The patent dates on the tubes give an indication of the earliest possible date the 3-weight tall clock could have been made.

Durfee's tubes, manufactured by US Tubular Bell Co., are 1-1/2" in diameter and range in length from approximately 40-1/2" to 59", with the hour tube being 70-1/2" long. The actual length may vary, because each tube is factory hand tuned to give it its best tone. In the process of tuning, the lower end

Figure 4.26. "Theodore Starr" stamped on nameplate.

Figure 4.27. "Tiffany & Co." stamped on nameplate.

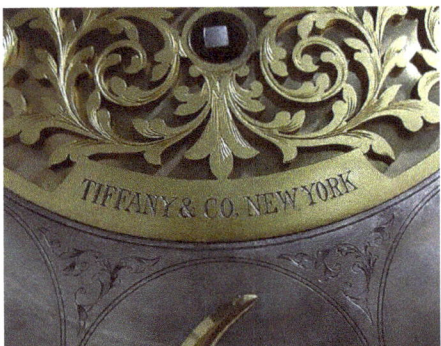

Figure 4.28. "Tiffany & Co. New York" stamped on nameplate.

Figure 4.29. "Walther H. Durfee" engraved on a rare embossed dial.

Figure 4.30. Stamped name and patent dates on tubes.

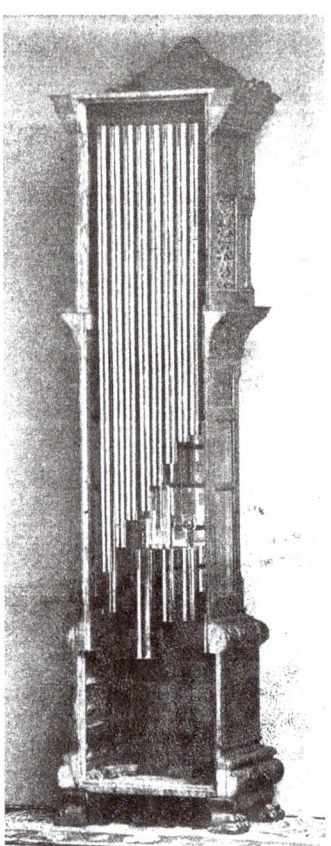

Figure 4.30A. Rear view of tubes hanging in case. See Figure 61.

of the tube is reduced in length until it produces the exact sound needed. The tubes are hung in the rear of the case (Figure 4.30A). The chiming tubes are suspended from an iron frame. When facing the clock, the smallest tube will be on the left side of the case and the longest tube on the right side (Figure 4.31). The hour tube is suspended from a separate iron frame mounted on the right side of the case (Figure 4.32). It is fair to assume the two iron mounting frames were either supplied by the tube maker or by a source close to Providence, RI. From 1887 to 1892 the hour strike hammer had a leaf spring striking arm with a string adjustment. The same string adjusting method was used on the chiming tube hammers (see arrow in Figure 4.32). Around 1892 the hour tube string adjusting method was replaced by a patented fine tuning adjustable hammer (Figure 4.33). The adjustable hammer has an English patent, for "Pat'd Feb. 28, 1888" is stamped on the assembly. Our research has been unable to determine who the actual inventor was or who was issued the patent. It was a much needed improvement because it now allowed the hammer head to be moved in or out to increase or decrease the striking force. This allowed the owner to adjust the hammer impact to create a more pleasing sound. The new adjustable hour hammer required a design change to the iron hour tube hanging frame.[4.12]

Figure 4.31. Front view of hanging tubes and chime hammers.

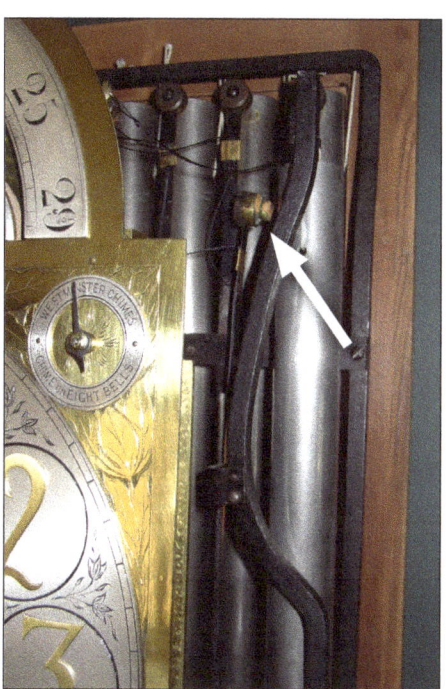

Figure 4.32. Hour strike hammer using string as adjustment (see arrow).

Figure 4.33. Patented fine tuning adjustment hour hammer.

Figure 4.34, right. Chiming tube clock's weights and brass bob.

Weights

After 1888 Durfee purchased his brass weights in the United States. It didn't take long for Durfee to figure out that he could save a considerable amount of money on overseas shipping cost, especially on heavy brass weights. Durfee found that he could have the complete brass weights manufactured locally. All Durfee had to do was copy or duplicate the weights on the 3-weight tall clocks he had received from England in prior years. He simply purchased the brass shells, brass bottoms, and brass caps locally, and polished and lacquered all brass parts. He also purchased the lead weights, by the size and poundage that were required. Then he simply assembled the components. We know that the weight assembly took place in Providence, RI, because the weight assembler used the local newspaper when packaging the lead weight and the brass shells. They wrapped the lead weights with newspaper to keep the lead weights from rattling in the brass shell.

Durfee may have used more than one brass shell supplier, because the shell sizes vary by 1/2" on the diameter and as much as 1" or 2" in their length. As a result of these variations, we have chosen to use only the median or average sizes of the finished brass shells, as well as the average poundage of the weights. The following dimensions are given from left to right, as the weights hang in the case (Figure 4.34):

- Strike weight—2-1/2" diameter and 11"—15-1/2 lbs.
- Time weight—2-1/2" diameter and 9-1/2"—13-1/2 lbs.
- Chime weight—3-1/2" diameter and 12-1/2"—40 lbs.

Pendulums

The 3-weight chiming tall clocks used four different style pendulums. The most common is a 6" diameter brass pendulum bob filled with lead. One-vial, two-vial, and three-vial mercury pendulum bobs were also used.

The brass pendulum bob (Figure 4.34) was used in at least 90 percent of the Durfee 3-weight tall clocks.

The next most popular bob was the one-vial mercury bob (Figure 4.35), which was used as early as 1888. The one-vial mercury bob is approximately 6-1/2" long and contained 9-1/2 lbs. of mercury (Figure 4.35A). Only a small number of the two-vial and three-vial pendulum bobs have surfaced (Figures 4.36 and 4.37). When customers wanted a more accurate timekeeper or when they wanted the beauty of a swinging glass mercury jar, they would purchase a mercury pendulum. A mercury pendulum bob could add $100-$250 to the base price of a new Durfee clock.

Pendulum Rods

Durfee mainly used wooden pendulum rods from 1887 to 1894 (Figure 4.37A), but examples show he used a few steel pendulum rods as early as 1890. By 1895 a great number of the J. J. Elliott movements being supplied to Durfee had steel pendulum rods. Many people refer to these steel rods as Invar rods, which is questionable. Invar is a nickel and iron alloy with an extremely low coefficient of expansion, meaning that any pendulum rod made of Invar could minimize any changes in timekeeping due to temperature changes. Because Invar was not invented until 1897, by Dr. Charles Guillaume,[4,13] we can easily conclude that Elliott and Durfee could not have used Invar until sometime after 1897.

Pendulum Bob and Rod Stamps

Many of Durfee's 3-weight tall clocks have matching numbers stamped on the pendulum bob and pendulum rod. In some cases these matching numbers are on the

Figure 4.35. One-vial mercury pendulum bob.

Figure 4.35A. Close-up of one-vial mercury bob.

Figure 4.36. Two-vial mercury pendulum bob.

Figure 4.37. Three-vial mercury pendulum bob.

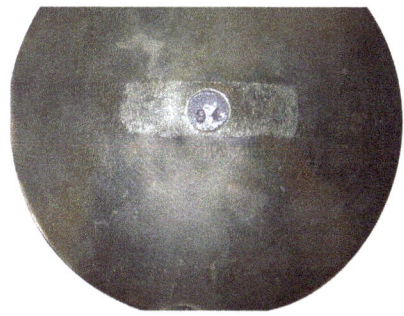

Figure 4.37A, left. Durfee's wooden pendulum rod "6794".

Figure 4.38, center. Brass pendulum bob "6794" identification stamp.

Figure 4.39, right. Pendulum rod "6794" identification stamp.

Figure 4.40. The "6794" identification stamp on seat board.

back of the dial plate, the back of the moondial, and, in some cases, on the movement plate. If the numbers are on the back of the pendulum bob (Figure 4.38) and at the top of the pendulum rod (Figure 4.39), these numbers must match. When stamped numbers are also on the dial plate, moondial, and the movement, they too must always match the bob and rod numbers. This matching number system was developed to keep all components for the J. J. Elliott movement intact so they could be easily identified at assembly time. In some cases the seat board was stamped with the matching number (Figure 4.40). In Figures 4.38 to 4.40, the number "94" was stamped on everything, but the number "67" shown in front of the "94" was only on the seat board. The number "67" could have stood for many things, such as the movement lot number, the movement model number, or the customer identification. We may never know the correct answer.

Finials and Pediments

Durfee used many different styles of finials and carved tops, depending on the model. In most cases Durfee used only wood finials when the case had a wooden column with wooden capitals and bases. The same was true when Durfee used brass finials, because the column's capitals and bases would be made of brass. Durfee's hood tops always had some type of detailed carving or special design. In all situations Durfee's finials and hood tops added grace and quality to the case. The following photographs, taken from Durfee's 3-weight clock cases shown in Chapter 3, are self-explanatory and little description is needed:

- Pattern 2 has three wooden hand-carved flame and urn finials (Figure 4.41).
- Pattern 4 has three wooden urn-shaped finials with the large center finial being a thing of beauty (Figure 4.42).
- Pattern 5 has two outer urn-shaped finials with a massive removable carved hood top (Figure 4.43).
- Pattern 11 has a removable crest carved top with the figure portraying the North Wind (Figure 4.44).
- Pattern 13 has three wooden finials with the hood top featuring a carved Viking head, located between the two swan necks, surrounded by a detailed carved floral motif (Figure 4.45).
- Pattern 15 has a single hand-carved acorn-style finial between the two carved swan necks (Figure 4.46).
- Pattern 18 has three wooden urn finials, wooden capitals and bases, and two wooden carved swan's neck rosettes (Figures 4.47 and 4.48).
- Pattern 18-V1 has three brass finials, brass capitals and bases, and two stamped brass swan's neck rosettes (Figure 4.49).
- Pattern 20 has a massive removable top with a large Satanic carved figure looking down on the clock winder (Figure 4.50).

Feet

Durfee's tall clocks always had feet of some sort; the base of the clock never rested directly on the floor. Over half of the 3-weight tall clocks had some type of carved claw feet. Examples of the feet Durfee casemakers used are shown in Figures 4.51 to 4.56.

Case Hardware

Durfee used only a few different escutcheons on the 3-weight tall clocks. The most common and popular escutcheon was the cast-brass lion head (Figure 4.57). The next most common escutcheon was the stamped-brass floral and ribbon design (Figure 4.58). Another style escutcheon, used on a few of the latter models, is the cast-brass facial design (Figure 4.59). Some of the model cases did not require or have room for an escutcheon and it was omitted.

Durfee always seemed to go the extra mile to try to incorporate top quality in everything. The waist door hinges on some cases are a good example. The higher price clocks have these cast brass engraved hinges (Figure 4.60). The normal case hinges are a standard plain flat brass hinge. The actual style hinges used varied from model to model. Many models used the plain flat brass hinges and the fancy hinges.

Sales Agents and Advertisements

Durfee did not waste any time introducing his 3-weight tall clock to the public through advertising. By early 1888, less than four months after producing his first chiming hall clock, he published an advertisement in *Harper's Magazine* announcing that he alone controlled all "United States Rights of Harrington's Patent Tubular Bells" (Figure 4.61). The ad included the following:

- The clocks were listed as "English Hall Clocks." All future ads dropped the word "English" and called them "Hall Clocks."
- By mid-1888 Durfee had established eight sales agents and pointed out that these "English Hall Clocks" could be seen in the agent's showroom. These early agents were located in New York (2), Philadelphia (2), Providence, Detroit, Kansas City, and St. Louis.
- This 1888 ad is the only one where Tiffany & Co., New York, is listed as an agent for Durfee.
- This was the only ad that showed an address for the Walter H. Durfee & Co. at "295 High Street, Providence, R.I." Future Durfee ads would only show "Walter H. Durfee & Co., Providence, RI."

Figure 4.41. Pattern 2, flame finials.

Figure 4.42. Pattern 4, urn finials.

Figure 4.43. Pattern 5, top and finials.

Figure 4.44. Pattern 11, carved top.

Figure 4.45. Pattern 13, finials and detailed carvings.

Figure 4.46. Pattern 15, acorn finial.

Figure 4.47. Pattern 18, wooden urn finials.

Figure 4.48. Pattern 18, wooden flame finials.

Figure 4.49. Pattern 18-V1, brass finials.

Figure 4.50. Pattern 20, carved top.

Figure 4.51. Durfee's most popular carved claw foot.

Figure 4.52. Large claw foot on top of a ball.

Figure 4.53. Three-toe claw foot.

Figure 4.54. Four-toe claw foot.

Figure 4.55. Winged angel foot.

Figure 4.56. Art deco-style foot.

- This ad identified the company as "Manufacturers and Importers of English Hall Clocks."
- The clock shown in this ad was never featured in Herschede's 1895 catalog. In Chapter 3 we assigned Pattern 9 to this clock. This Pattern 9 was never shown again in any ads or on his stationery letterhead.

The next known advertisement was published in May 1890 and disclosed a drastic change in the Walter H. Durfee & Co. method of operation that changed the future direction of the company.

Durfee's 1890 ad states, "In buying a hall clock, as much care should be taken in selecting a case that is well made, and making sure that the cabinet-work is first class, as you would take in securing a movement of the best quality." Durfee's ad then adds, "Never buy an imported case. It will not stand the variations of our climate." This confirms several conversations we had with the late Dana Blackwell, in which he stated, "Elisha Durfee, Walter's nephew and partner, told him a number of times that Durfee purchased his cases from American furniture makers." Elisha even gave Blackwell the names of two well-known Boston cabinetmaking firms who had made cases for Walter H. Durfee & Co.: Turner and Co. and L. E. Kimball and Co. on Decatur Street. Boston is 50 miles from Providence, which would allow Durfee to visit their shops quite often and be personally involved in the case designs. There may have been others, such as Breitenstein & Sons, Providence, RI, who did major antique repair and cabinetwork for Durfee for over 25 years. One of the Breitenstein ads even illustrates a Durfee 2-weight tall clock. This 1890 ad disclosure verifies that Durfee was now purchasing his cases solely in the United States, not England.

Figure 4.57. Lion's head escutcheon.

Figure 4.58. Stamped brass escutcheon.

Figure 4.59. Cast brass facial design escutcheon.

Figure 4.60. Cast brass engraved waist door hinges.

By 1890, with the clock cases now being manufactured in the United States, Durfee was only purchasing and importing his movements and dials in England. All remaining components were coming from US suppliers.

The 1890 ad also states, "These Tubular Chiming Clocks are fully protected by Letters [of] Patent, and persons purchasing should see that every bell is stamped with our name." The ad further states, "They [chiming clocks that do not have Durfee tubes] that are not sold by our authorized sells Agents, are an infringement of our rights and are subject to heavy royalty [fines] on either the buyer or seller. Clocks imported with Tubular Bells come under this infringement, as we control all U.S. Rights." It became apparent that Durfee's tubular bell patent rights were being challenged by others. Durfee faced this challenge by clock importers and makers for the next 12 years, right up to his court case *Durfee vs. Bawo*, in October 1902, when the US Circuit Court ruled that his rights to restrict the use of tubular bells in clocks had expired. Many Durfee ads from 1890 to 1896 stated these tube disclosure conditions, but to no avail.

Durfee worked to increase his sales outlets. By 1890 Durfee had 14 agents, by 1891 he had 16, by 1892 he had 24, and by 1896, his last known ad, his agent list had declined to 21. As early as 1890 Durfee ads stated, "Each agent had an album containing photographs of all of our

Figure 4.61. 1888 Durfee ad. WALTER H. DURFEE CO.'S *BREAKFAST CALL CHIMES CATALOG*, CA. 1896.

Table 4.1.

DURFEE'S ADS SUMMARY CHART

(Year That Ad Appeared) AD YEAR		1888	1889	1890	1890	1891	1892	1893	1896
(Pattern Number of Clock in Ad) PATTERN NO.		#10	#4	#4	#4	#22	#22	#4	#18
(Number of Agents Shown in Ad) AGENTS		8	13	13	14	16	24	24	21
Agent	**City**								
Bailey, Banks & Biddle (PA)	Philadelphia	X	X	X	X	X	X	X	X
Bradstreet, Thurber & Co. (MN)	Minneapolis		X	X	X	X			
Bunde & Upmeyer (WI)	Milwaukee						X	X	X
Caldwell (James E. Caldwell & Co) (PA)	Philadelphia	X	X	X	X	X	X	X	X
Cowell & Hubbard Co. (OH)	Cleveland		X	X	X	X	X	X	X
Grogan (J. C. Grogan) (PA)	Pittsburgh		X	X	X	X	X	X	X
Hennegen Bates & Co. (MD)	Baltimore						X	X	X
Herschede (Frank Herschede) (OH)	Cincinnati					X	X	X	X
Hudson (J. B. Hudson) (MN)	Minneapolis						X	X	X
Jaccard Watch & Jewelry Co. (KS)	Kansas City	X	X	X	X	X	X	X	
Mermod & Jaccard Jewelry Co. (MO)	St. Louis	X	X	X	X	X	X	X	X
Moores & Winder, Troy (NY)	Troy						X	X	
Nathan, Dohrmann & Co. (CA)	San Francisco								X
Parker & Davis (CT)	Bridgeport						X	X	X
Robbins (F. A. Robbins) (MA)	Pittsfield						X	X	X
Rodgers & Pottinger (KY)	Louisville					X			
Saxton (W. H. Saxton) (CT)	New London						X	X	
Scooler (M. Scooler) (LA)	New Orleans						X	X	
Seymour (Jos. Seymour & Sons (NY)	Syracuse						X	X	
Smith (Green Smith Watch Co.) (CO)	Denver						X	X	X
Spauling & Co. (IL)	Chicago		X	X	X	X	X	X	X
Starr (Theodore B. Starr) (NY)	New York	X	X	X	X	X	X	X	X
Tiffany & Co. (NY)	New York	X							
Tilden, Thurber & Co. (RI)	Providence	X	X	X	X	X	X	X	X
Turner (Henry A Turner & Co.) (MA)	Boston		X	X	X	X	X	X	X
Walk (Bingham & Walk) (IN)	Indianapolis					X			
Walk (Julius C. Walk) (IN)	Indianapolis						X	X	X
Woods & Hosley, Springfield (IL)	Springfield			X	X	X	X	X	X
Wright, Kay & Co. (MI)	Detroit	X	X	X	X	X	X	X	X
Total Ad Agents Listed	**29**								

Figure 4.62. Clock in agent's album.

designs." If only one of those albums could be found, many questions about Durfee's actual clock production could be answered. An example of one of the clocks in the agent's album, Pattern 20, is shown in Figure 4.62.

In Durfee's last known hall clock ad, published in 1896, Durfee states, "We have also just finished seven new designs in Hall Clocks, in Colonial, Henry II, and Louis XVI styles" (Figure 4.63). Why did Durfee design seven new hall clocks? Because none of the agent's albums, mentioned earlier, have been found, it is impossible to know what these seven new designs looked like or if any of them were made.

A summary chart, compiled from Durfee's eight published ads from 1888 through 1896, shows the 29 agents who sold his clocks (Table 4.1).

Breakfast Chimes

The 1896 ad also introduces his "Breakfast Calls Chimes" manufactured by US Tubular Bell Co., of which he was then president. These chimes were made in sets of two, three, and four tubes. The tubes chord or tone being the first and third for the two-tube; the first, third, and fifth for the three-tube; and the first, third, fifth, and eighth for the four-tube, of the octave in D (Figure 4.64).[4.14] The tubes are 1-1/2" in diameter and range in length from 3' to 5'; are heavily nickel plated; are suspended on oxidized silver brackets of ornamental design; and are mounted on highly finished carved shields in solid mahogany and oak of various patterns (Figure 4.65). The tube assembly also came with a specially constructed hammer that produces the purest and sweetest tube tones (Figure 4.66). The tubes are stamped, "Walter H. Durfee" with the four patent dates. At the very top of the tube is the stamped word "STRIKE" in an oval, signifying the location where the hammer should strike the tube.

Figure 4.63. Durfee's 1896 ad.

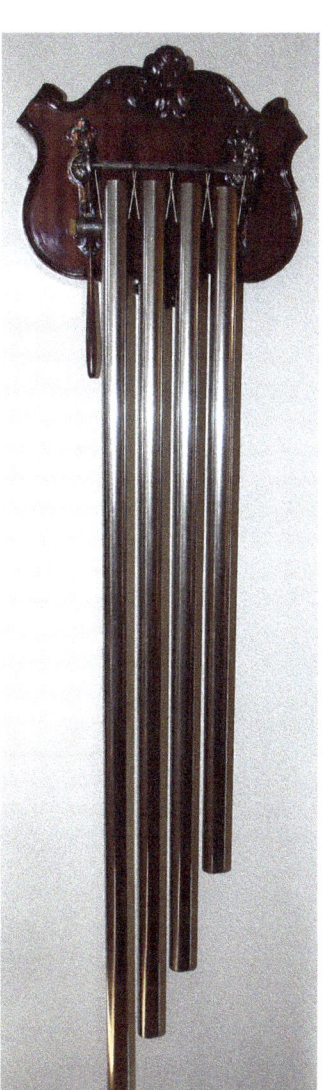

Figure 4.65. Four-tube set of breakfast chimes.

Figure 4.66. Close-up of hammer and shield.

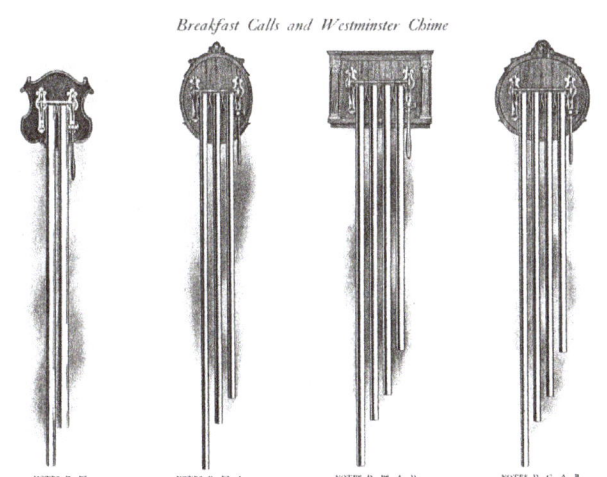

Figure 4.64. Four sets of the breakfast chimes.

Walter H. Durfee & His Clocks

Chapter 5

Curtis Girandole Clocks

Walter Durfee has become synonymous with the majestic grandfather clocks he produced from 1883 to roughly 1910; however, very little is known about his reproduction of the Lemuel Curtis girandole. While we were researching Durfee in the late 1970s and early '80s, we communicated with Durfee's sister-in-law, Hope Holdcamper, who was in her nineties at the time. She had recalled Durfee telling her that he had made six Curtis reproductions, of which he was extremely proud of them. By December 1981 we had only located one Durfee reproduction of a Curtis girandole (Figure 5.1). Then in 2008 another Durfee Curtis girandole reproduction appeared at an auction on Cape Cod, MA. Since then, we have found three more, bringing the total to five.

Durfee was an expert and authority on Lemuel Curtis. In the December 1923 issue of *The Magazine Antiques,* he wrote the article "The Clocks of Lemuel Curtis," which covers Curtis's life, his relationship with J. L. Dunning, and the clocks and movements he made and used.[5.1] In the article were photographs of eight original Curtis clocks, of which two were girandoles. From statements in the article it appears that Durfee had access to these original Curtis clocks, giving him the opportunity to examine them in detail. This is especially true of the two girandoles, because he described their different features in great detail. Durfee noted that the Curtis clocks were rare, and because of their beauty and rarity they commanded high prices, making "faking" profitable. Because of this he warned Curtis clock buyers to be cautious and to have them authenticated.

Durfee stated the following about the Curtis girandole in Figure 5.2:

> The case is of solid mahogany. The two front panels, the rope turning, the eagle ornament and base, and the base bracket are pine, ornamented with gold-leaf; and the bezel is of brass. The side brasses are cast brass, cored in at the large end and drilled at both ends for accommodating the stamped brass rosettes. The handles of the catches of the top and

Figure 5.1. Durfee's first Curtis girandole reproduction, ca. 1908. COURTESY OF PETE VANDER POEL.

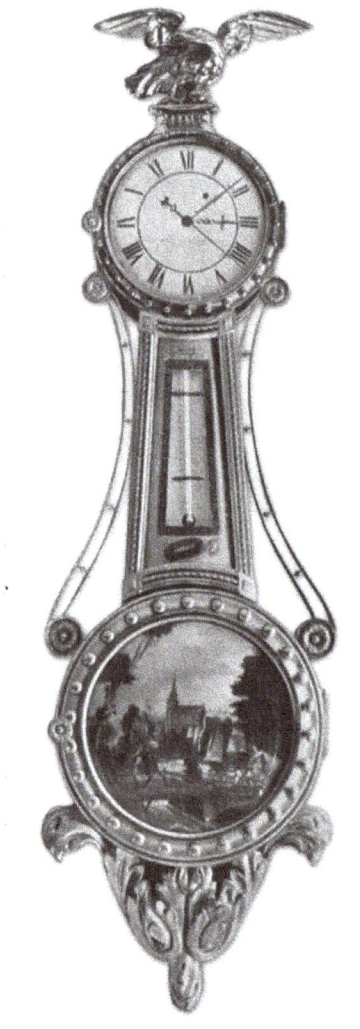

Figure 5.2. Original Lemuel Curtis girandole. THE MAGAZINE ANTIQUES, DECEMBER 1923.

bottom doors are finished off with a similar rosette of smaller pattern. The corners of the center panel are ornamented with another stamping. On the bezel door there are twenty-seven balls, but if we count the place where the catch takes the place of a ball, there would be twenty-eight. These balls are made of brass. On the lower door there are twenty-six balls. The clock has the usual Curtis hands, which consist of a series of loops or circles. On this clock the hour hand has four loops and the minute hand five … . The clock has, moreover, two unusual and striking characteristics: a sweep second hand, and a thermometer in the center panel . . . with the word Patent is painted on the glass . . . The painting in the lower glass represents *Paul Revere's Ride*, with the Old North Church in the background."[5.2]

In 1908 Durfee produced his first Curtis girandole reproduction (Figure 5.1). In general it is a copy of the Curtis girandole featured in his 1923 article (Figure 5.2) and described in detail above. The major variations in the Durfee reproduction are as follows:

- Durfee did not use Curtis's fancy door catches but replaced them with standard banjo side catches.
- He used a painted porcelain dial with his name "Walter H. Durfee Providence" painted on the dial above the center arbor opening (Figure 5.3).
- The center panel is all glass because he chose not to duplicate the thermometer Curtis used in his original. The center glass has a different reverse painting and the word "Patent" painted at the bottom of the glass. Each corner of the frame has a small brass parallelogram-shaped ornamental stamping (Figure 5.4).
- Curtis's four rosette brass stampings on the two side arms were replaced with four round painted porcelain inserts (Figure 5.4).
- Unable to obtain an original Curtis movement, Durfee chose to use a Waltham banjo movement stamped "5665" (Figure 5.5).

Figure 5.3. Durfee's signature name on the porcelain dial. COURTESY OF PETE VANDER POEL.

Figure 5.4. Center view of girandole showing waist glass, side arm porcelain inserts, and corner brass ornamental stampings. COURTESY OF PETE VANDER POEL.

Figure 5.5. Waltham movement No. 5665 used in Durfee's first girandole. COURTESY OF PETE VANDER POEL.

- For future identification of the true clockmaker Durfee stamped "Walter H. Durfee Providence" on the inside of the lower case in a semi-circle (Figure 5.6).
- The reverse painting on the lower glass happens to be an exact copy of the painting *Paul Revere's Ride*, with the Old North Church in the background (Figure 5.7). We can assume that Durfee and the unknown artist had access to the original Curtis girandole in order to duplicate the original Curtis glass as closely as they did (Figure 5.2).

During the construction of the Curtis girandole reproduction, Durfee created a mystery. Written in ink inside the center throat glass frame (Figure 5.8) is:

"W. D. Coggeshall
Presented by Mr. Durfee
March 1908
W. H. Durfee maker"

Also on the lower inside surface of the throat glass reverse painting, written in pencil, is "#1" and "W. D. Coggeshall March 1908." These findings raise questions about Coggeshall's identity and the reason Durfee presented this Curtis girandole to him. A genealogical study found that W. D. Coggeshall was an older cousin to Walter H. Durfee. His full name was Walter Durfee Coggeshall Jr.; he was born in the Providence, RI, area in 1850, seven years before Durfee. Coggeshall and his wife Josephine Harriett Coggeshall, born July 15, 1848, in Saratoga, NY, moved from Providence to London, England, around 1880. Although they retained their US citizenship, they maintained residency in London for the rest of their lives. In 1887 they had a daughter Mary Josephine Coggeshall, who became an English citizen. In 1912 she married a man with a surname of Home and became Mary Josephine Home. On Walter D. Coggeshall's many trips to the United States, passage information listed him as a merchant.

Records show that Durfee made at least 11 transatlantic round trips to England from 1883 to 1903. Durfee stayed 20 to 45 days in England on each trip. His last known trip to England was on April 20, 1903, just six months after the New York Circuit Court ruled, on October 4, 1902, in the *Durfee vs. Bawo* case, that Durfee's sole patent rights to the Harrington chime tubes had expired. Up to this time all American chiming hall clock manufacturers had to purchase their tubes, as well as the expensive Elliott chiming clock movements, from Durfee.

On December 29, 1902, Herschede Hall Clock Co. was incorporated and immediately started to make its own chiming tubes and importing less expensive German movements for its hall clocks. Other hall clock manufacturers, like Waltham, Colonial, and Borgfeldt, quickly followed suit. As a result of this court ruling Durfee's hall clock business quickly deteriorated, and his trips to London to purchase supplies and the Elliott movements from his English suppliers came to an end. Because Durfee no longer made these trips, he had to have an agent or representative living in London to handle his remaining

Figure 5.6. Durfee's ID stamp on lower part of case. COURTESY OF PETE VANDER POEL.

Figure 5.7. Lower tablet is a copy of Paul Revere's Ride. COURTESY OF PETE VANDER POEL.

Figure 5.8. "W. D. Coggeshall" name in ink on inside of waist glass frame. COURTESY OF PETE VANDER POEL.

clock business. Who better to handle this than his cousin Coggeshall?

Other transatlantic crossing records show that Coggeshall made at least eight round trips with the last four after Durfee's final London trip in 1903. Whether it was coincidental or not, Coggeshall's last crossing was in 1921, shortly after Durfee closed his manufacturing shop at 121 Pond St. in Providence and went into semi-retirement. As a result of Durfee's working relationship with Coggeshall or being a frequent guest in the Coggeshall home while he was in London, Durfee may have believed he owed something to Coggeshall. Durfee probably selected his Curtis girandole reproduction No. 1 (see Figure 5.1) as a gift of appreciation for Coggeshall's help and generosity. We do know that the clock remained in London with the Coggeshall family until the mid-1960s when an American antique dealer purchased it, brought it back to the United States, and later sold it to a clock collector in California. At the time the clock was purchased in London, Mary Josephine Home, the only child of Walter D. Coggeshall, was in her seventies and the clock could have been part of her estate.

We believe that when Durfee made his Coggeshall Curtis girandole in 1908 he did not intend to make another. But 10-12 years later he made six more. By that time he was an expert and authority on Lemuel Curtis and his clocks. Durfee believed this next group of Curtis girandole reproductions would be his legacy to his banjo clock-manufacturing business. Because Curtis made variations in his girandole design, Durfee had to decide which original to copy or whether to create his own designs by picking the best characteristics of the Curtis girandoles he had studied. Durfee chose the latter.

The following excerpt is from Durfee's 1923 article showing his in-depth knowledge of the Curtis girandoles:

> They vary greatly in minor details, but all have the same general characteristics. The base bracket, of an acanthus pattern, is virtually identical in all cases. The eagle finial ornament may be either the spread-wing pattern illustrated, with the eagle turning toward the right, or a drop-wing pattern with the eagle facing the front. The side brasses do not vary.
>
> The dials are always of iron slightly convex and may be unmarked, or marked L. Curtis; L. Curtis, Patent; Warranted by L. Curtis, or Curtis & Dunning. They are never marked with Dunning's name alone. When dials are unmarked, the name Curtis should appear in the center glass. The name Curtis or the word Patent is never painted on the glass of the lower door as it is in some of the Willard clocks. Generally the dials have Roman numerals, as is the case with most of Curtis's clocks; but occasionally the Arabic system is used. Generally, too, the dials are undecorated: but sometimes they are ornamented with gold-leaf, a sun-burst pattern around the center arbor and keyhole, a scalloped edge of black and gold, or a thin gold line around the dial on the inside of the circle of the numerals.
>
> The lower door, together with the balls on it, is always made of wood. The balls may be of two sizes. When the larger size is used, the number is twenty six; when the smaller size is used, the number varies from thirty-five to thirty-seven. It is generally thirty-five. The bezel door is generally made of brass, rarely of wood. When this door is of brass, the balls should be of a similar material; but if it is made of wood, the balls should be of wood also. The balls on this door vary very little in size in different examples and their number should be twenty-eight or twenty-nine.[5.3]

Durfee's detailed research provided him with guidelines on how to design and make his final Curtis reproductions.

Variation I

Around 1920, Durfee produced his second girandole (Figure 5.9) and made many design changes from his first, including the following:

- The number of balls on the lower bezel went from 26 to 36 (Figure 5.10). The number of balls on the upper bezel remained the same at 28.
- The lower tablet became *Perry's Victory on Lake Erie* (Figure 5.10) in place of *Paul Revere's Ride*. Part of Oliver Hazard Perry's famous dispatch to General Harrison is inscribed on the glass: "We have met the enemy and they are ours." This *Perry's Victory* scene is a copy of the Benjamin B. Curtis tablet painting on one of L. Curtis's original girandoles, which is signed on the back "Painted by Benj. B. Curtis."[5.4] Benjamin B. Curtis was Lemuel's brother and was well known for his high-quality workmanship in painting clock tablets and ornamental gilding. The lower tablet is an exact duplication, because the artist for Durfee's reproduction girandole was Daniel J. Steele, who happened to own the original Curtis girandole. Steele of Milton, MA, was one of the most talented early twentieth-century ornamental glass and dial painters.[5.5] Many of his reverse painted glasses can be found on Durfee and Waltham banjo clocks.
- The waist glass had a new design and the four small brass stampings on the outside corners of the waist glass frame were omitted (Figure 5.11A). The word "Patent" was replaced with the artist name, "D. J. Steele" (Figure 5.11B).
- The four porcelain landscape discs on the two side arms were eliminated (Figure 5.11A).
- The dial artist, Steele, copied the dial of his original Curtis girandole[5.6] and replaced the Arabic numerals with Roman numerals (Figure

5.12A). A painted enameled dial replaced the porcelain dial. The gold ring inside the hour numbers was omitted. The L. Curtis gold sunbursts were added around the wind arbor and center arbor holes. A gold-painted ring was added to the outside edge of the dial. The location of Durfee's name moved from above the center arbor to below it (Figure 5.12B).

- The "Walter H. Durfee" ID stamp on the front inside location on the case was removed and replaced with two "Walter H. Durfee" round ID stamps on the inside face of the lower backboard similar to those shown in Figure 5.18.

Figure 5.11A. Durfee's Variation I girandole, new design of waist glass. COURTESY OF THE STANLEY WEISS COLLECTION.

Figure 5.9. Durfee's Variation I girandole, ca. 1920. COURTESY OF THE STANLEY WEISS COLLECTION.

Figure 5.10. Durfee's Variation I girandole, lower door and *Perry's Victory* tablet. COURTESY OF THE STANLEY WEISS COLLECTION.

Figure 5.11B. Durfee's Variation I girandole, artist name, "D. J. Steele", painted on lower part of waist glass. COURTESY OF THE STANLEY WEISS COLLECTION.

Figure 5.12A. Original Curtis girandole dial owned by D. J. Steele showing Arabic numbers that Durfee copied. *THE MAGAZINE ANTIQUES*, DECEMBER 1923.

Figure 5.12B. Durfee's Variation I, dial. COURTESY OF THE STANLEY WEISS COLLECTION.

- The factory added "Waltham Clock Co." to the front plate of the Waltham banjo movement No. 6340 (Figure 5.13).
- It has standard banjo hands in place of the Curtis looped hands. These could be replacement hands.
- The gilded eagle and the acanthus base remained the same.
- A brass strip pendulum tie-down was added. Assuming that Hope Holdcamper, Walter Durfee's sister-in-law, is correct—that Durfee made six Curtis girandole reproductions—it appears that Durfee only made one Variation I clock, or two at the most.

Variation II

In 1922 Durfee made his last production runs and slightly altered the previous design to produce his second variation (Figure 5.14). Changes consist of the following:

- The sunbursts around wind hole and center arbor openings were retained, while a gold band below the numbers was added to the dial.
- A loose brass ring replaced the gold-painted outside edge of the dial.
- The hands reverted to the original Curtis famous four-loop hour hand and five-loop minute hand (Figure 5.15).
- The waist glass reverse painting was redesigned with more color and the artist's name was omitted. A small round brass stamping was added to the corners of the frame (Figure 5.16).
- The movement manufacturer was changed from Waltham Clock Co. to E. Howard & Co. Boston (Figure 5.17). This appears to have been a special order with Howard. The nickel-plated movements are numbered No. 2 to No. 5. These movements are similar to the Howard No. 5 but they are not a true Howard No. 5. A "Curtis-style" pendulum tie-down was added (Figure 5.18). The Howard movement has a shorter pendulum length than the Waltham movement used in the previous girandoles. (Compare Figures 5.18 and 5.6.)

Figure 5.13. Early Variation II girandoles used Waltham movements, such as the No. 6340 shown above.

Figure 5.14. Durfee's Variation II Curtis girandole, ca. 1922.
COURTESY OF GEORGE TARDIFF.

Figure 5.15. Variation II dial and hands. COURTESY OF A. SUE WEISER.

Figure 5.16. Variation II waist glass design. COURTESY OF GEORGE TARDIFF.

Figure 5.17. Later Variation II girandoles used E. Howard movements as shown above. COURTESY OF GEORGE TARDIFF.

Figure 5.18. Lower view of case showing shorter pendulum length, ID stamps, tie-down, and level for setting clock up. COURTESY OF GEORGE TARDIFF.

Walter H. Durfee—Chapter 5 • 79

- The Waltham movements were mounted directly to a spacer board attached to the inside of the backboard; the Howard movements are mounted to a brass plate that is attached directly to the backboard (Figure 5.19). Note that Durfee's ID number "6" is stamped on the case inside the brass mounting plate (Figure 5.19) and the Howard movement and pendulum used with this mounting plate are stamped number "5" (Figure 5.17). The movement numbers are not necessarily the same as the case number.
- The movement weight (Figure 5.20) is shaped more like a Howard weight than a normal banjo weight. Even though it is a Howard weight, it still has the normal Durfee round ID stamp, which he used on all his banjo weights.
- A brass level marked "Stanley New Britain, Conn. USA" at the bottom of the case was added for setting up the clock (Figure 5.20). The lower tablets remained the *Perry's Victory* and again were painted by D. J. Steele. The glasses were signed on the inside "Painted by D. J. Steele 1922" on the first group of glasses on this model (Figure 5.21A) and just plain "D. J. Steele 1922" on the second group (Figure 5.21B). The eagle and acanthus bracket remain the same (Figures 5.22 and 5.23).

Figure 5.19. Brass movement bracket for mounting movement. COURTESY OF GEORGE TARDIFF.

Figure 5.20. Howard movement weight with Durfee's ID stamp. COURTESY OF GEORGE TARDIFF.

Figure 5.21A. Artist signed this lower tablet "Painted by D. J. Steele 1922" in the early part of 1922. COURTESY OF A. SUE WEISER.

Figure 5.21B, left. Artist signed this lower tablet "D. J. Steele 1922" in late 1922. COURTESY OF TOM JEPSEN.

Conclusion

We have seen or verified five Curtis girandole reproductions Durfee made, but we believe Durfee made seven at the most. The first one, the Coggeshall, was made in 1908; one Variation I was made in 1920; three Variation IIs were made in 1922. The two that are missing are either one more Variation I and one more Variation II or no more Variation I and two Variation IIs.

Over the years other reproduction clock manufacturers, such as Waltham Clock Co., Elmer Stennes, Ted Burleigh, Foster Campos, and Wayne Cline, have made beautiful Curtis girandole clocks, but none have come close to the quality or beauty of the Durfee girandole.

Figure 5.22. Eagle used on all Durfee's girandoles.

Figure 5.23. Acanthus lower bracket that was used on all Durfee's girandoles.

Examples of Banjo Clocks Made by Walter H. Durfee from 1908 through 1918

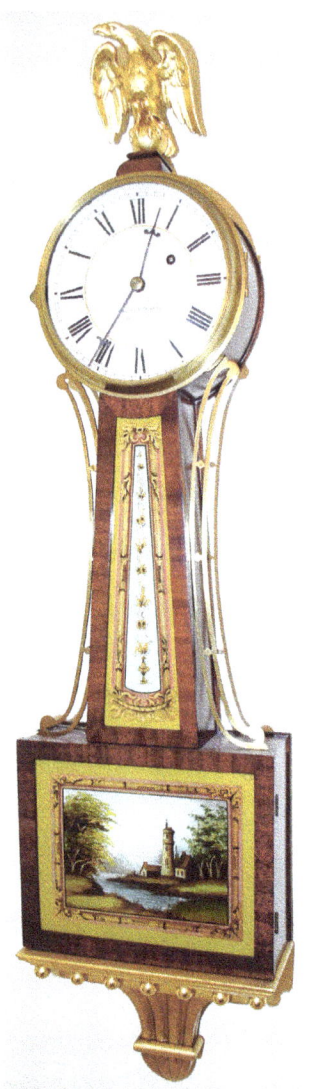   

Figure 5.24. Durfee's Willard-style banjo.

Figure 5.25. Willard-style banjo with reversed side arms.

Figure 5.26. Durfee's lyre banjo with full base.

Figure 5.27. Durfee's short base lyre banjo.

Figure 5.28. Example of the Waltham movement used in his banjo clocks.

NOTE: For more information on Durfee's banjo clocks see pages 10-12.

Walter H. Durfee & His Clocks

Chapter 6

Restoration of a 9 Tube Hall Clock

Figure 6.1. This Durfee Pattern No. 24 tall clock was manufactured by Walter H. Durfee, Providence, RI, ca. 1902-1904. It has a 103" tall mahogany case and a J. J. Elliott, London, England, tubular bell chime movement chiming on 9 hanging Harrington tubes located in the back of the case.

Dan Buffinga, a professional clock restorer by vocation, is a strong believer that any clock restoration work must be done by the book—without any shortcuts. This commitment manifested itself in the restoration of this one-of-a-kind Durfee Pattern No. 24, a 9 tube chiming hall clock owned by Mr. and Mrs. Joel T. "Bo" Cheatham III of Henderson, NC.

Buffinga, a friend, knew about my interest in Durfee clocks and informed me about this restoration project involving a model manufactured near the end of the Durfee era. He shared with me the 285 photographs he took of the restoration process.

Some of these images are included in the following photo essay that follows the work completed on the case, dial, movement, and chime mechanism. Although the case was in fair condition for its age, it had accumulated dust, smoke, grease, and layers of polish and wax over the past 110 years. The case required a complete cleaning and a hand rubout to restore its original finish. The silver on the dial had deteriorated to a black shade, requiring re-silvering of all the dial components. The movement hadn't been cleaned in at least 40-50 years and was not in running condition, so it required a complete overhaul. The chiming mechanism was in complete disarray and required a complete overhauling of the housing, cylinder, hammers, and timing components.

Case Restoration

Figure 6.2. An example of the dirt on the hood before any restoration or removal of the hood columns. The first step in restoring the case required complete disassembly of all case components.

Figure 6.3. After the hood columns were removed, dust from more than 100 years was found behind them.

Figure 6.4. Accumulation of dust and dirt was found behind the base of the waist columns.

84 • Chapter 6—Walter H. Durfee

Figure 6.5. Dust had collected on the hood-side sound panels and behind the front hood columns. Layers of dirt had accumulated on the flat surfaces. The sound cloth had rotted over the years and needed to be replaced.

Figure 6.6. The claw feet showed years of damage from the elements and physical contact.

Figure 6.7. The hood body, door, side sound panels, dial board, columns, and columns' carved capitals were restored and prepared for final assembly.

Figure 6.8. Side view of the hood after restoration of the hood's column assembly and the replacement of the sound cloth.

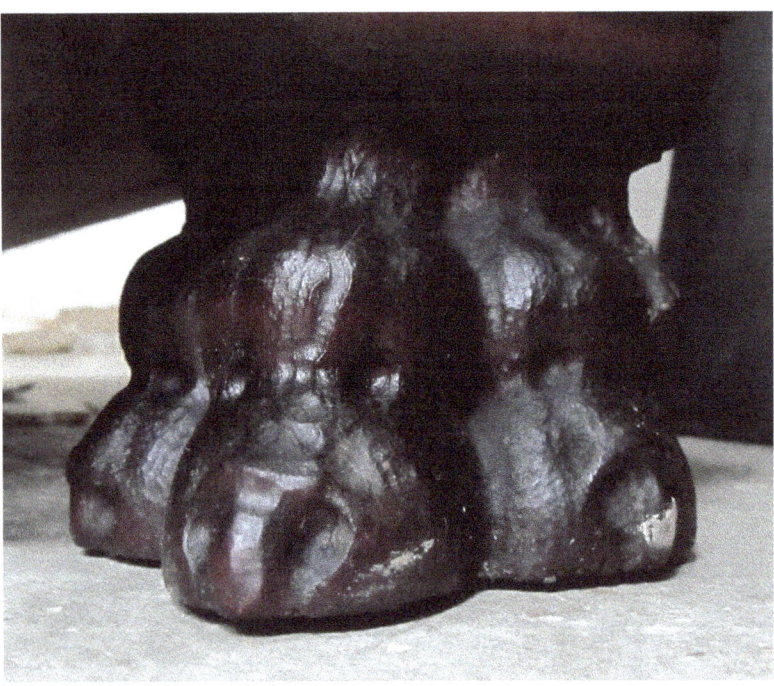

Figure 6.9. The carved claw foot was repaired of any damage, and all dirt and grime were removed.

Figure 6.10. Front view of the Durfee Pattern No. 24 case after its complete case was restored.

Figure 6.11. Hood view of restored case highlighting the delicate carving and hand-carved wood finials.

Figure 6.12. Waist view of the restored case showing its intricate detailed carvings, fluted waist columns, and an oval beveled door glass. At the time of this article's publication, this is the only known Durfee hall clock with an oval waist glass. Oval waist doors did not appear in hall clocks until ca. 1903.

Figure 6.13, right. Base view of the restored case displaying a beautiful carved shell with a flowing ribbon and flat fluted side columns.

Dial Restoration

Figure 6.14. Dial and movement assemblies, removed from the case, before any restoration. Because of heavy oxidation and surface wear, eight dial components needed complete re-silvering, the brass base plate needed to be polished and lacquered, and the moon dial disk need to be cleaned.

Figure 6.15. The dial unattached from the movement with the moon dial, hands, and numbers removed.

Figure 6.16. The finished brass dial plate after a thorough cleaning, polishing, and lacquering.

Figure 6.17, right. Eight completely re-silvered dial parts: the chapter ring, the second bit chapter ring, the chime chapter ring, the tune chapter ring, the moon dial date plate, the moon's two semi-sphere plates, and the jeweler's identification nameplate.

Figure 6.18, below. In the middle of the dial and above 6 o'clock is the identification plate of the jeweler who sold the clock, "Tiffany & Co New York". All of the dial's numbers retained their original gold plating and their luster after a run through the ultrasonic cleaner.

Figure 6.19. Back view of the dial before any restoration work after it had been detached from the movement. The little bent wires, pointed to with white arrows, hold the dial's gold numerals to the chapter ring.

Walter H. Durfee—Chapter 6 • **89**

Figure 6.20. Back view of the dial plate after it had been cleaned, polished, and restored.

Movement Restoration

Figure 6.21. Front view of the three-train movement before any cleaning and restoration. Note the movement manufacture's number "506" stamped on the lower center of the front plate. This number is also found on the pendulum rod.

90 • Chapter 6—Walter H. Durfee

Figure 6.22. Rear view of the movement before disassembly and restoration.

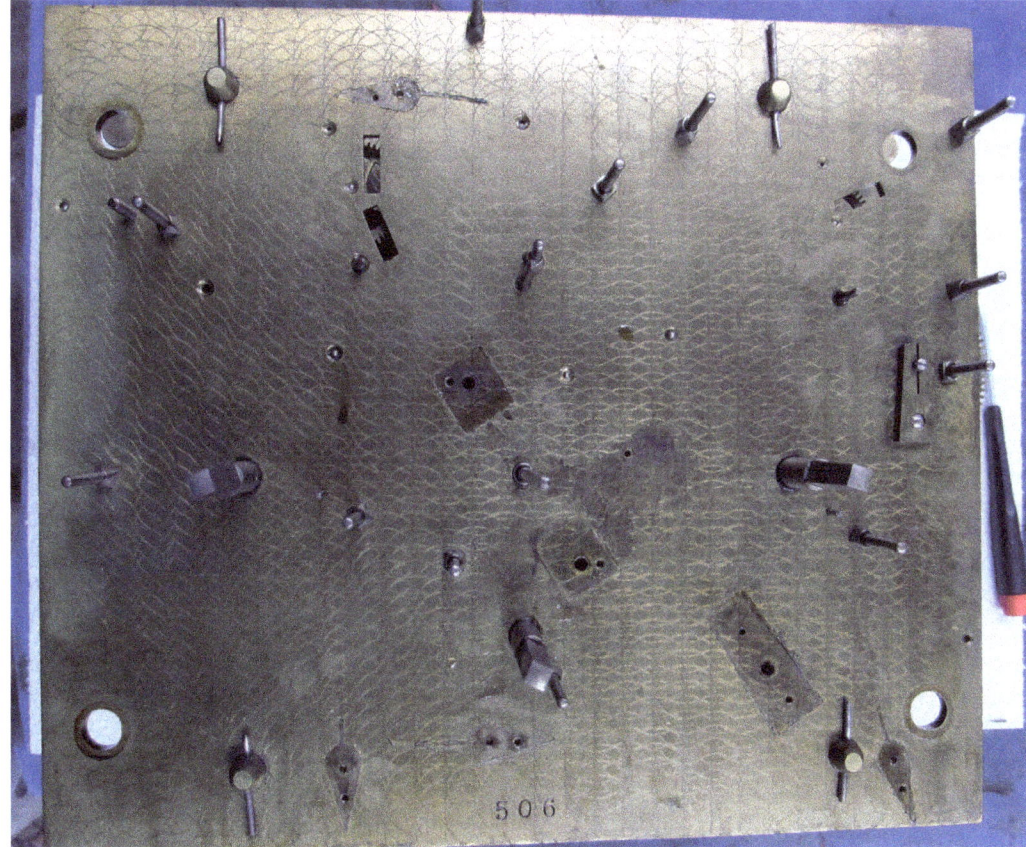

Figure 6.23. Front view of the movement after removal of all striking and motion levers and wheels. Before polishing the front plate, all of the levers' posts and pins shown here had to be removed.

Figure 6.24. The total accumulation of 34 levers, wheels, bridges, and assemblies that were removed from the front and rear plates before their cleaning. Each part had to be individually cleaned and polished before being placed back on the movement.

Figure 6.25. Top view of the movement's internal gearing before removing the front plate, which clearly illustrates the need for a thorough cleaning to remove all of the grease, oil, and grime accumulated over the past 50-60 years.

Figure 6.26. View of the movement's internal wheels and gears after the front plate was removed.

Figure 6.27. Examination of the movement plates showed that the pivot for the time 3rd wheel was badly worn and had to be repaired and the pivot hole, identified by the white arrow, had to be bushed.

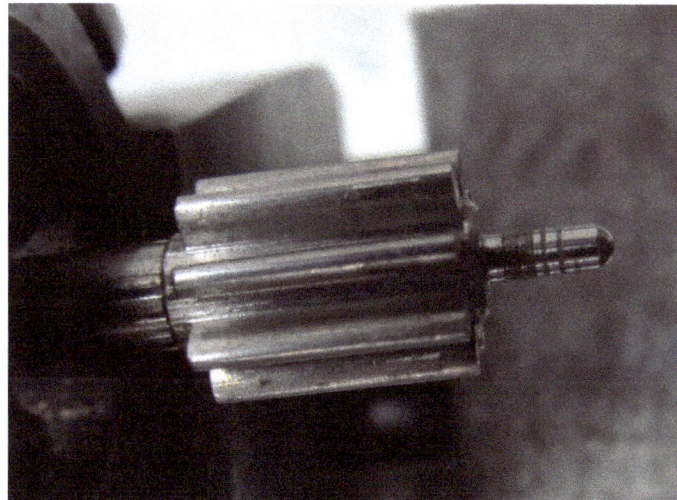

Figure 6.28. View of the worn 3rd wheel pivot with deep gouges in the pivot end before it was repaired.

Figure 6.29. The repair of the worn 3rd wheel is completed. With a smaller pivot the rear plate had to be bushed.

Walter H. Durfee—Chapter 6 • 93

Figure 6.30. The verge pallets were so badly pitted from wear, as indicated by the arrows, that they needed to be resurfaced and repaired.

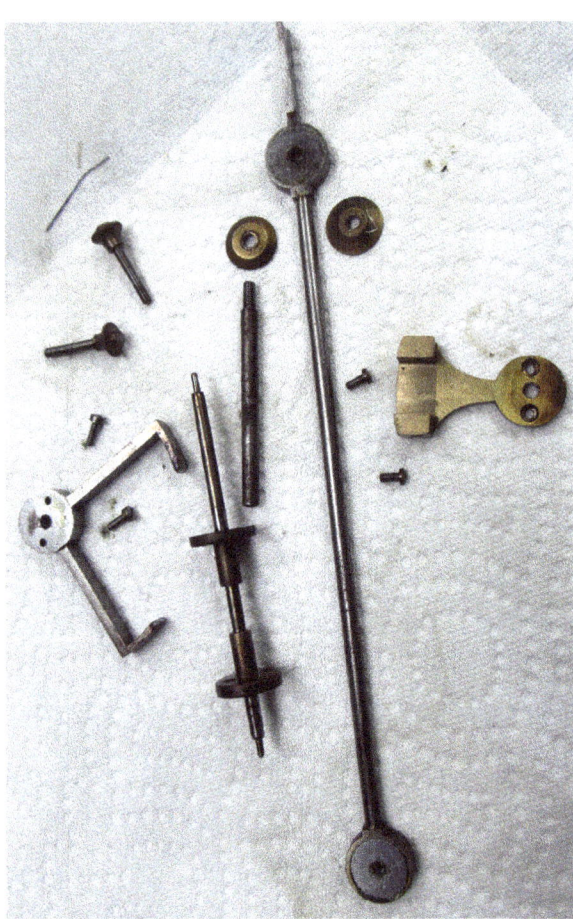

Figure 6.31. The entire verge and crutch assembly had to be completely disassembled before any cleaning and repairs could be made to the verge pallets.

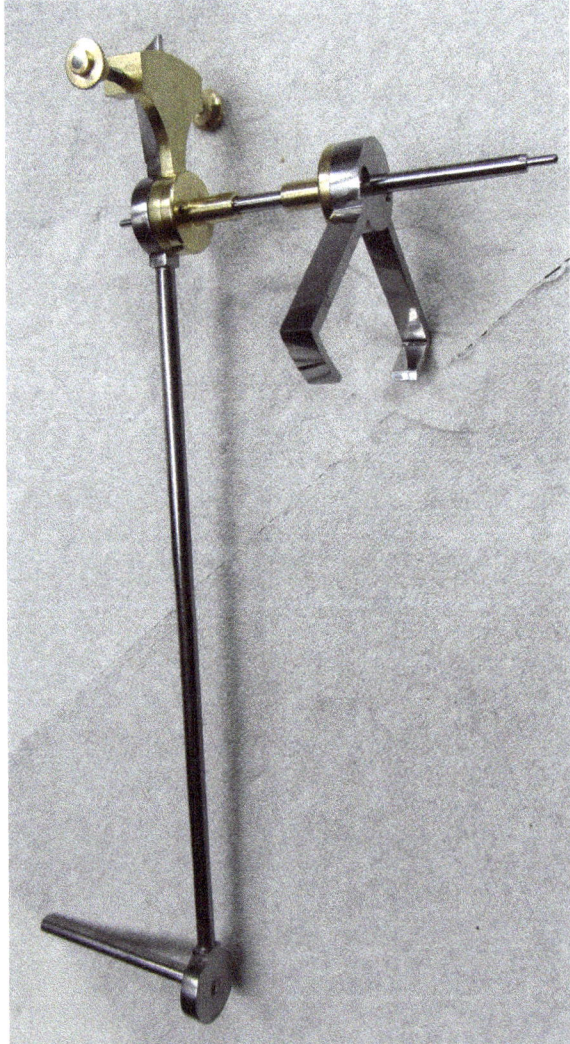

Figure 6.32. The verge and crutch assembly after all repairs and cleaning were completed.

Figure 6.33. Because the escape wheel assembly was heavily covered with grime, it needed a thorough cleaning and polishing before it could be placed into the movement.

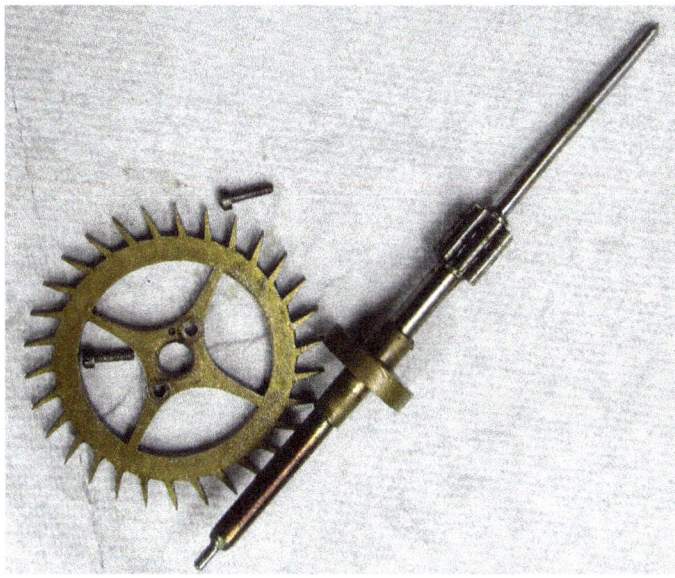

Figure 6.34. The escape wheel assembly had to be completely disassembled before any cleaning or repairs.

Figure 6.35. The finished escape wheel assembly after cleaning and ready for final assembly.

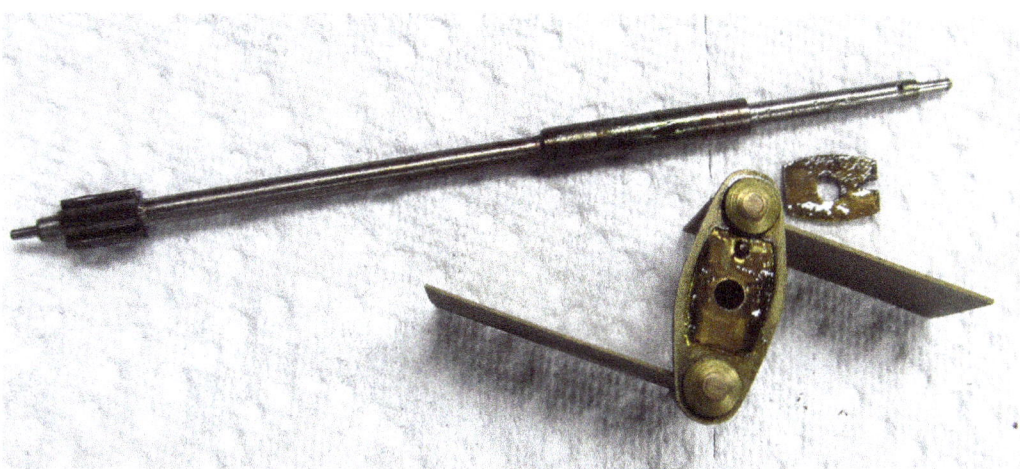

Figure 6.36. One of the two fly assemblies was covered with old oil and grime and required disassembly before cleaning.

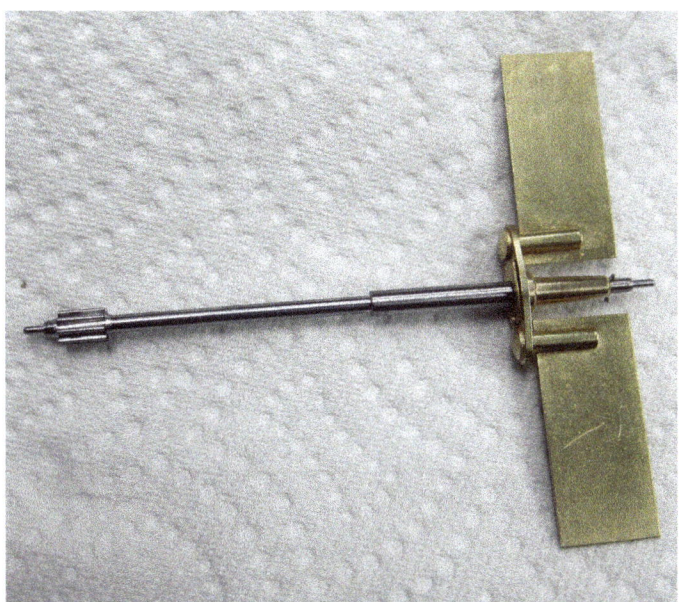

Figure 6.37. The fly assembly was restored and cleaned, and ready to be reattached to the movement.

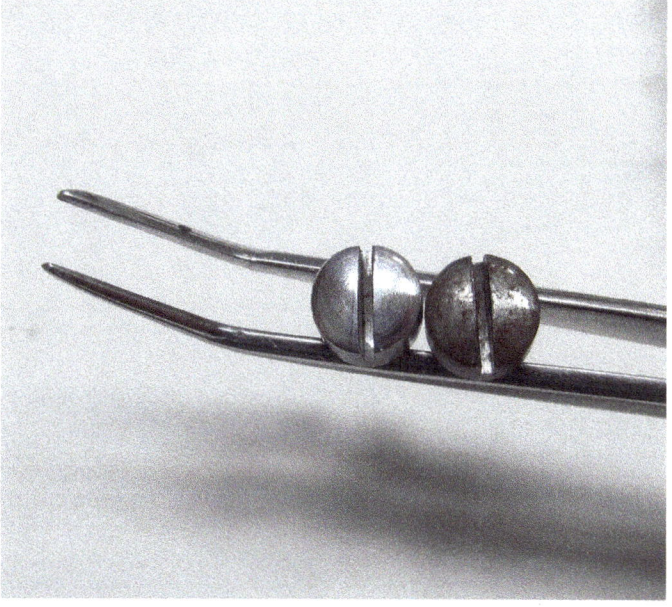

Figure 6.38. Often neglected by repairers are the screwheads. Extra time was taken to debur and polish them.

Figure 6.39. The clock movement's plates, parts, and components after all repairs and cleaning and ready for final assembly.

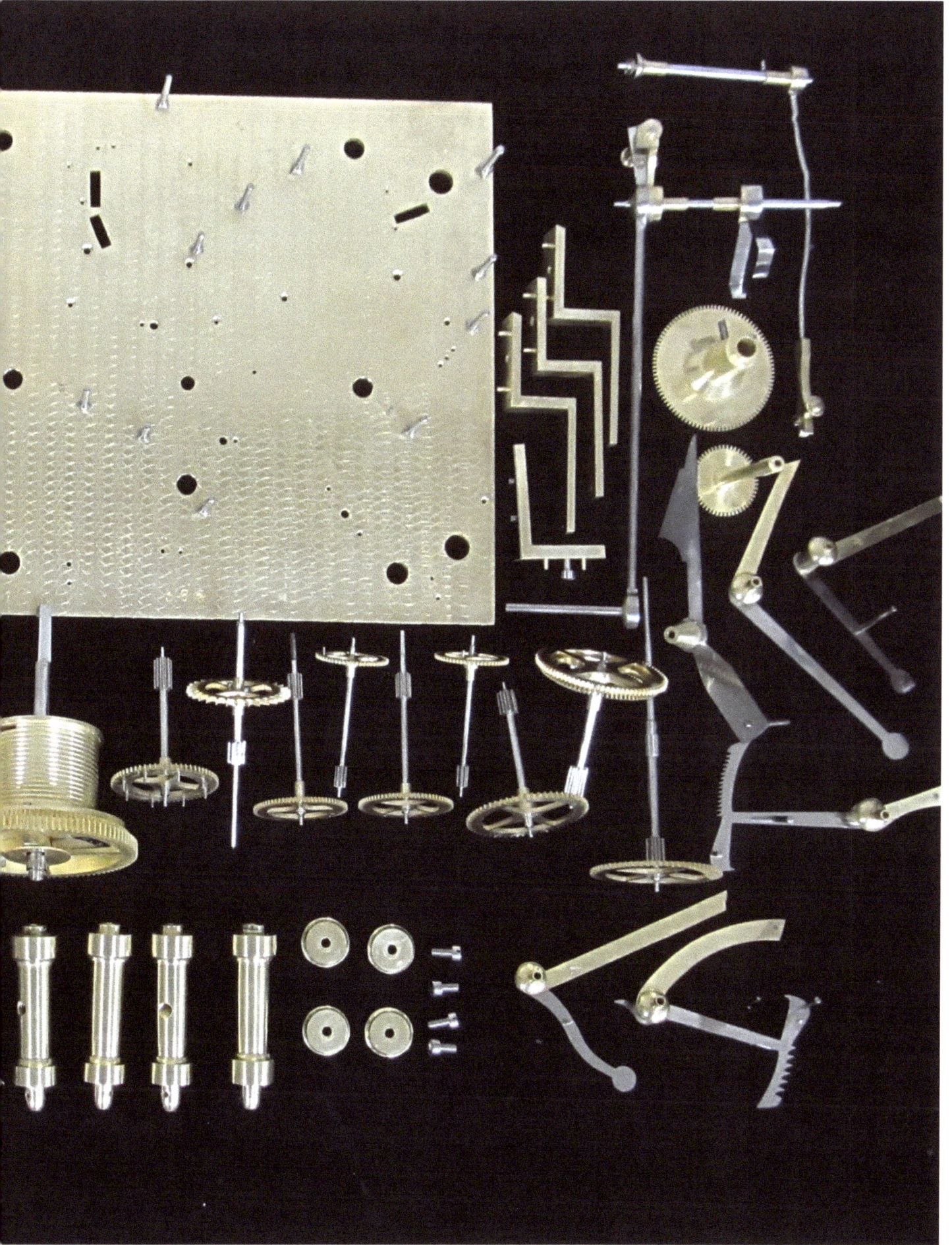

Figure 6.40. After the internal parts and gearing were cleaned and repaired, they were reassembled to the backplate. (Compare to before cleaning in Figure 6.26.)

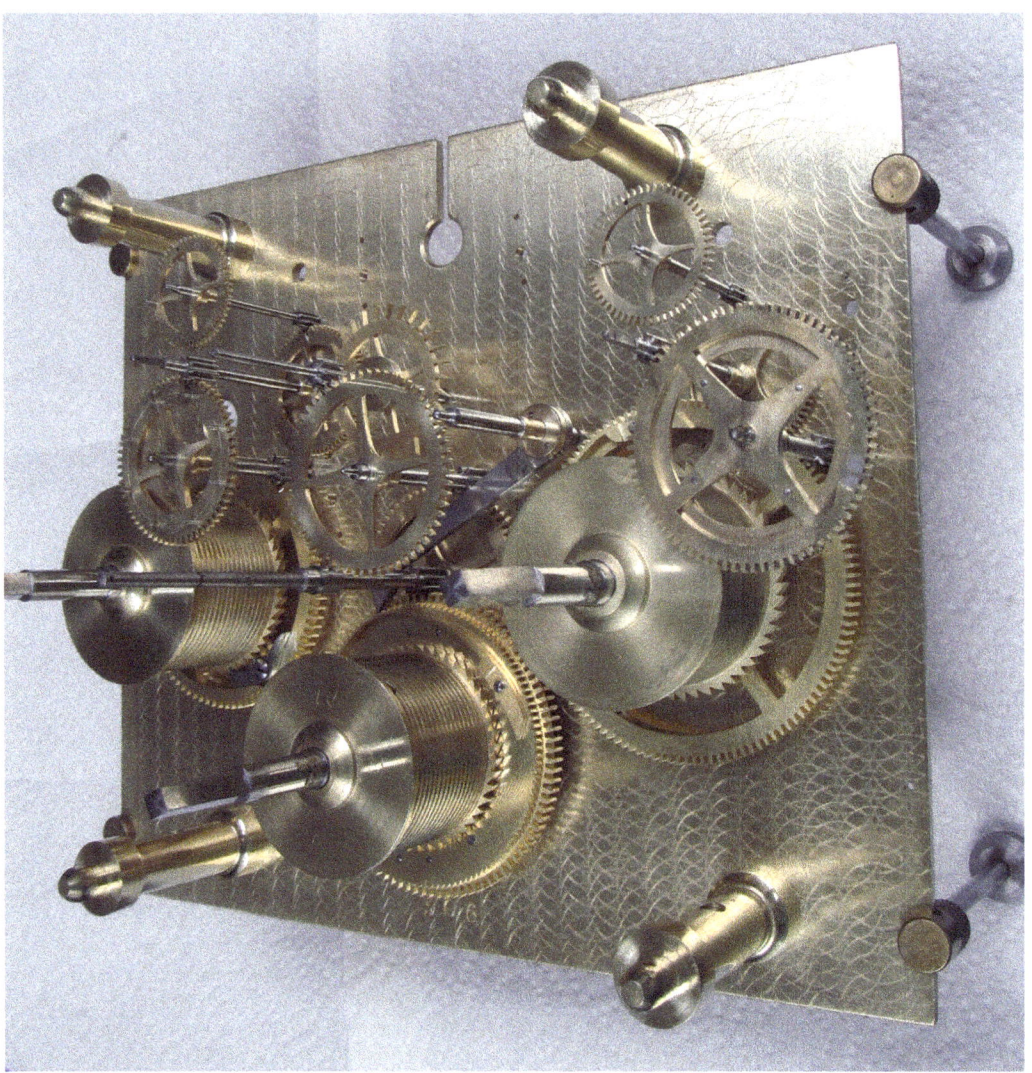

Figure 6.41. The cleaned and polished front plate being mounted to the movement. Before mounting, all of the posts and pillars were removed, polished, and reattached. (Compare to before cleaning in Figure 6.23.)

Figure 6.42. Top view of internal gearing after cleaning and polishing. (Compare to before cleaning in Figure 6.25.)

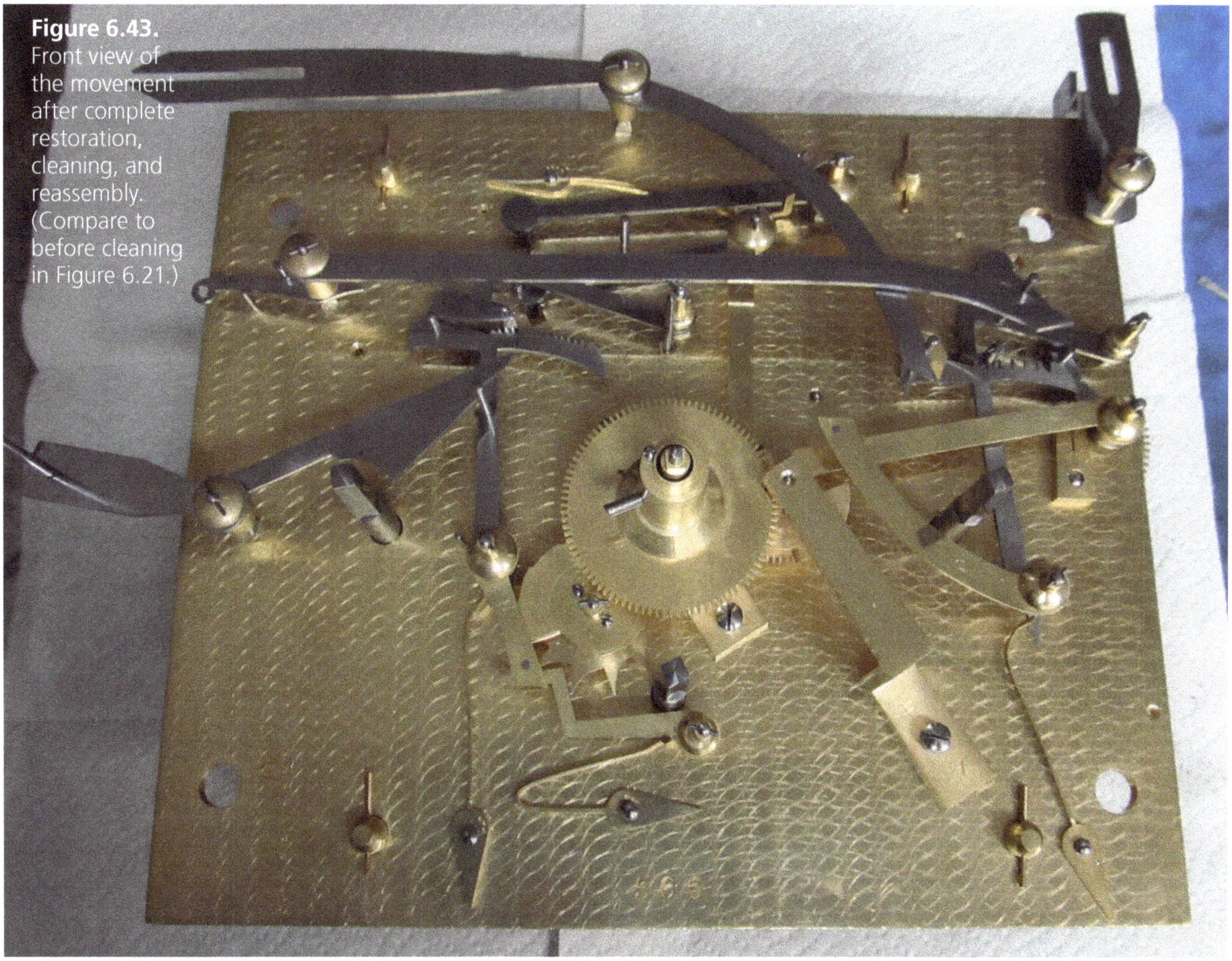

Figure 6.43. Front view of the movement after complete restoration, cleaning, and reassembly. (Compare to before cleaning in Figure 6.21.)

Figure 6.44. Rear view of the movement after complete restoration, cleaning, and reassembly. (Compare to before cleaning in Figure 6.21.)

Figure 6.45, below left. Right-side full view of the movement after restoration, cleaning, and reassembly.

Figure 6.46, below right. Left-side full view of the movement after restoration, cleaning, and reassembly.

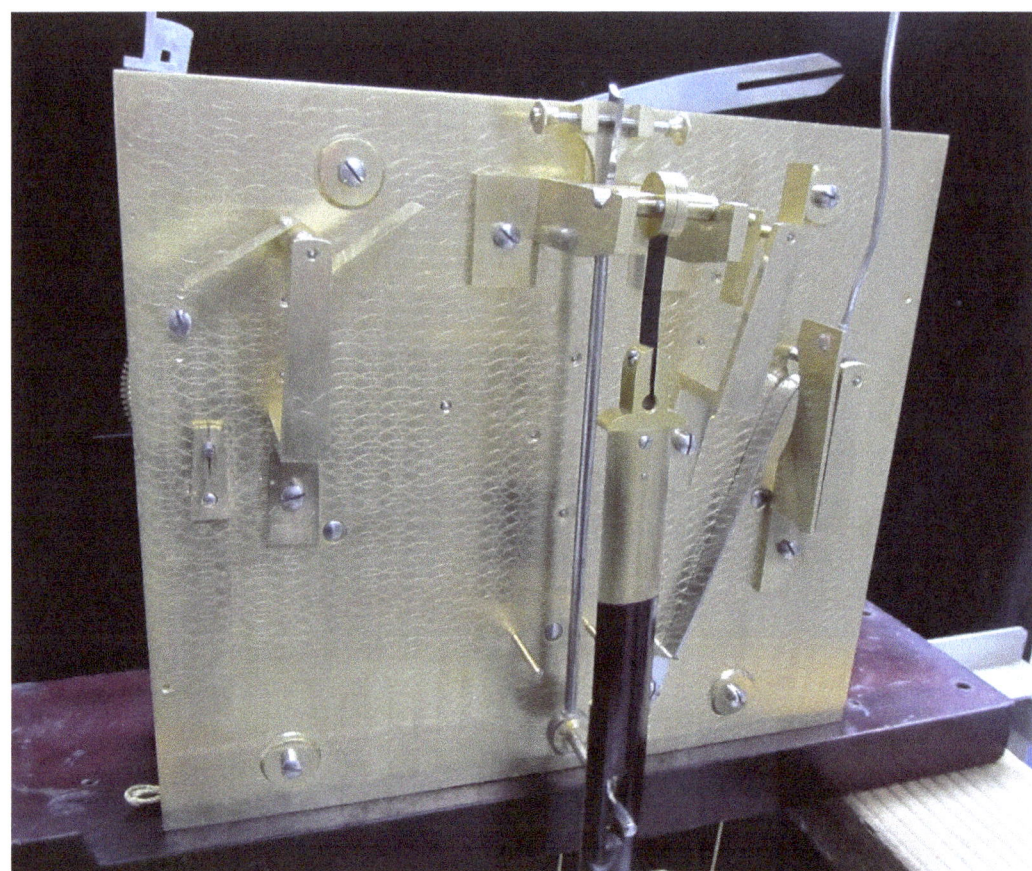

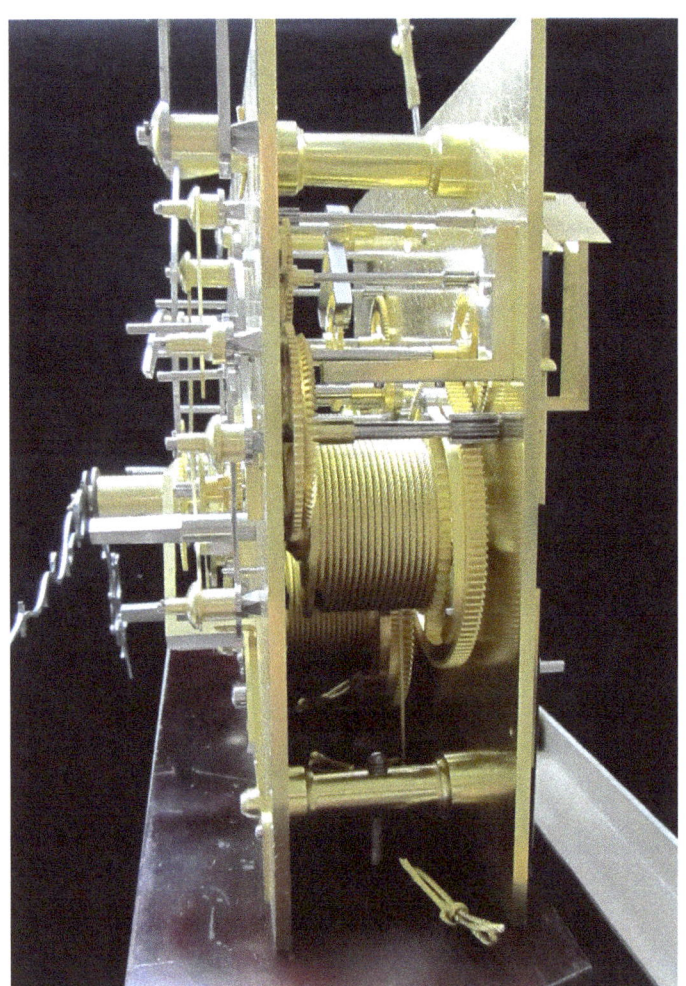

Figure 6.47. The pendulum bob needed to be polished after 100 years of neglect.

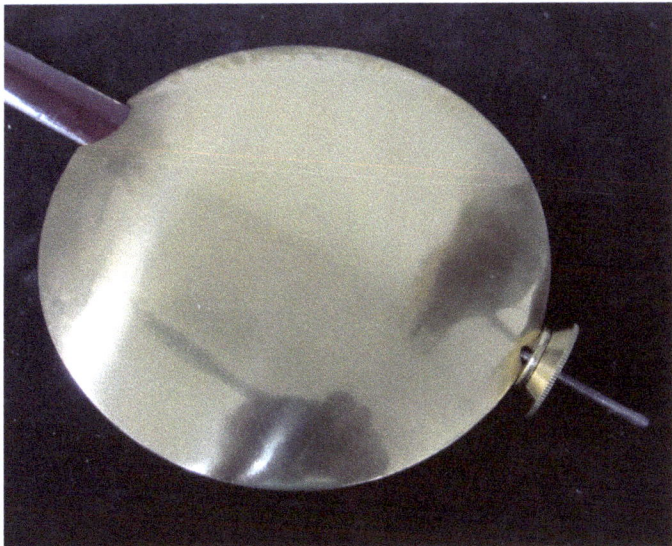

Figure 6.48. The finished brass bob after cleaning, polishing, and lacquering.

Figure 6.49, right. Although the pendulum rod and suspension assembly were in working order, they were cleaned and polished to go with the rest of the clock's restoration.

Figure 6.50, left. Close-up of the disassembled parts of the pendulum rod and suspension assembly before they were cleaned.

Figure 6.51, below left. The pendulum rod and the suspension assembly after being restored. Note the number "506" located below the suspension assembly and above the top of the pendulum rod, which matches the movement number on the front plate shown in Figures 6.21 and 6.23. Many movement makers supplied the movement and pendulum and stamped the corresponding number on the pendulum used for testing and setting up the clock and sold the two as a set.

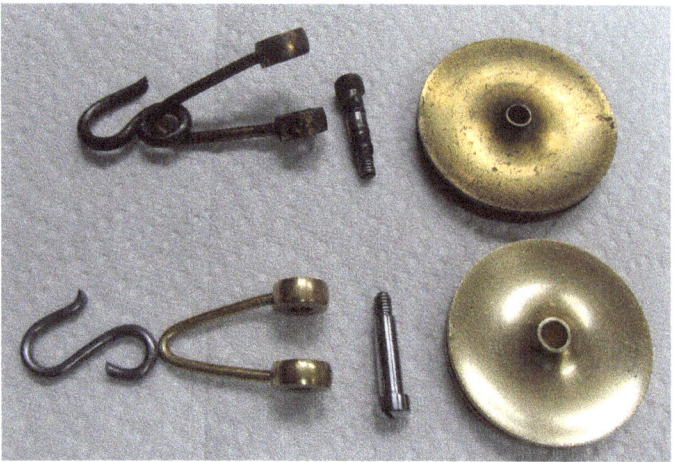

Figure 6.52. The movement pulleys before and after restoration.

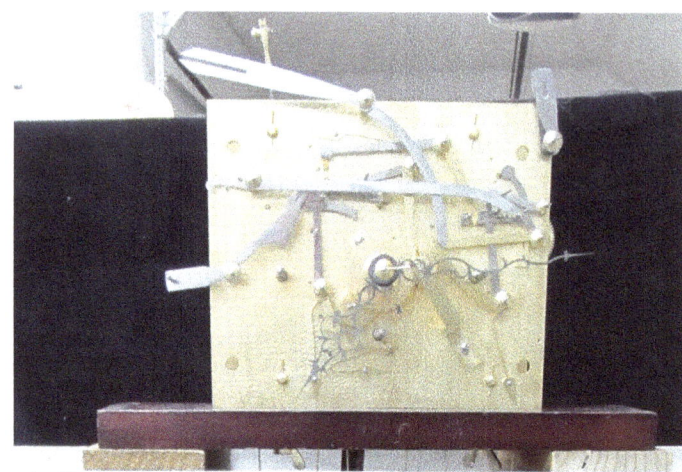

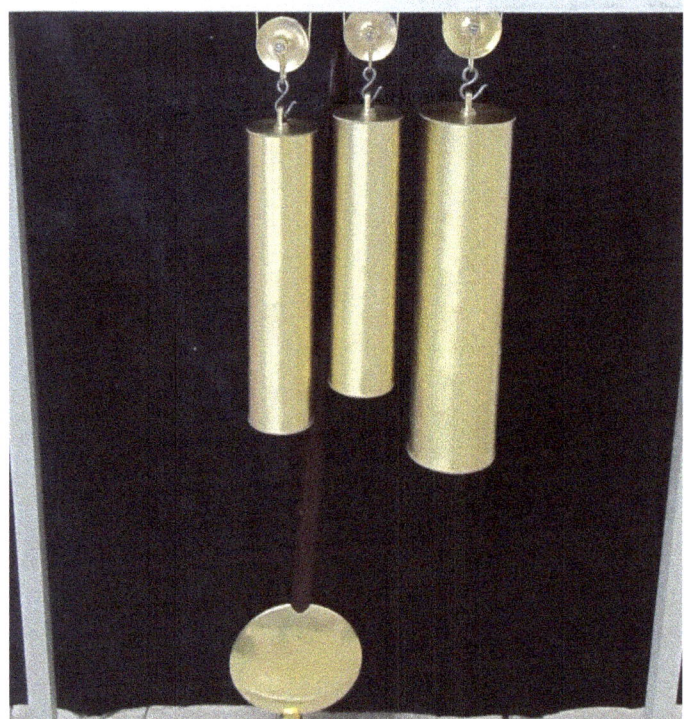

Figure 6.53, right. Three weights after cleaning, polishing, and lacquering. While disassembling the large chime weight, a copy of a Providence, RI, newspaper, dated February 20, 1901, was found wrapped around the lead weight, suggesting the approximate date the clock was manufactured or sold.

Figure 6.54. The original hands were so badly bent, broken, and soldered over the years that they had to be replaced. (See the replacement hands in Figure 6.18.)

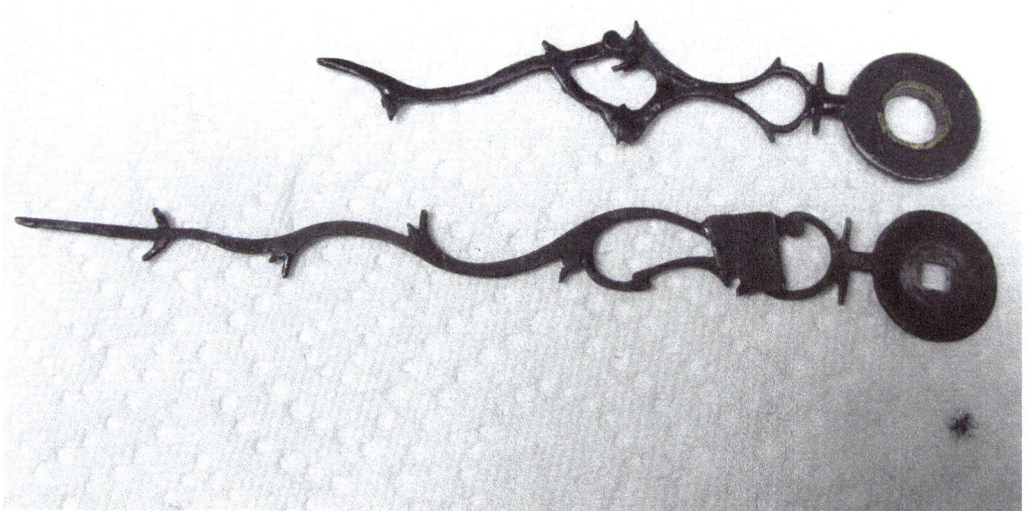

Chime Mechanism Restoration

Figure 6.55. The last major step in restoration of this clock was overhauling the chiming mechanism. Here is the cylinder assembly before any disassembly and cleaning. Note the deteriorated shock absorber cords that had to be replaced during the cylinder's final assembly.

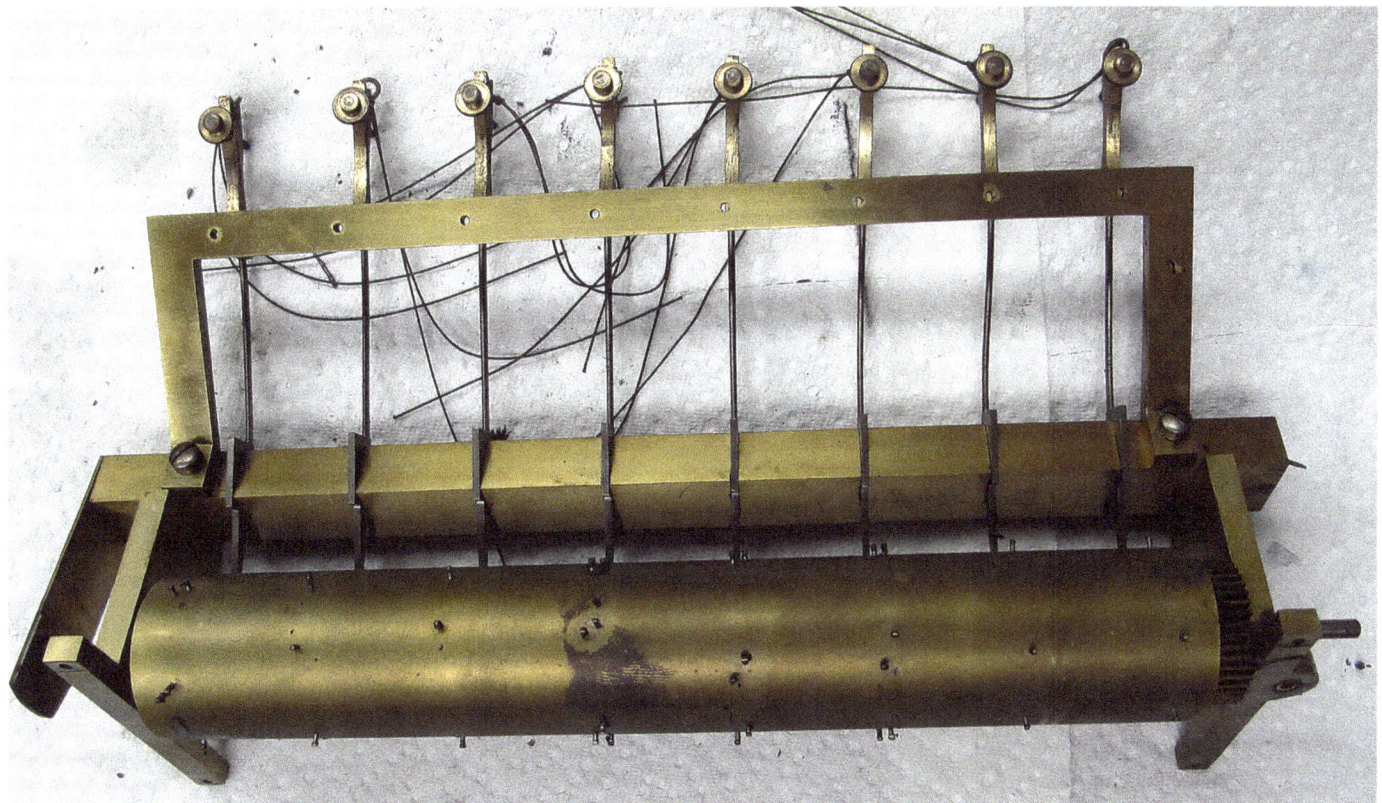

Figure 6.56. The chime cylinder assembly detached from the movement and awaiting full restoration.

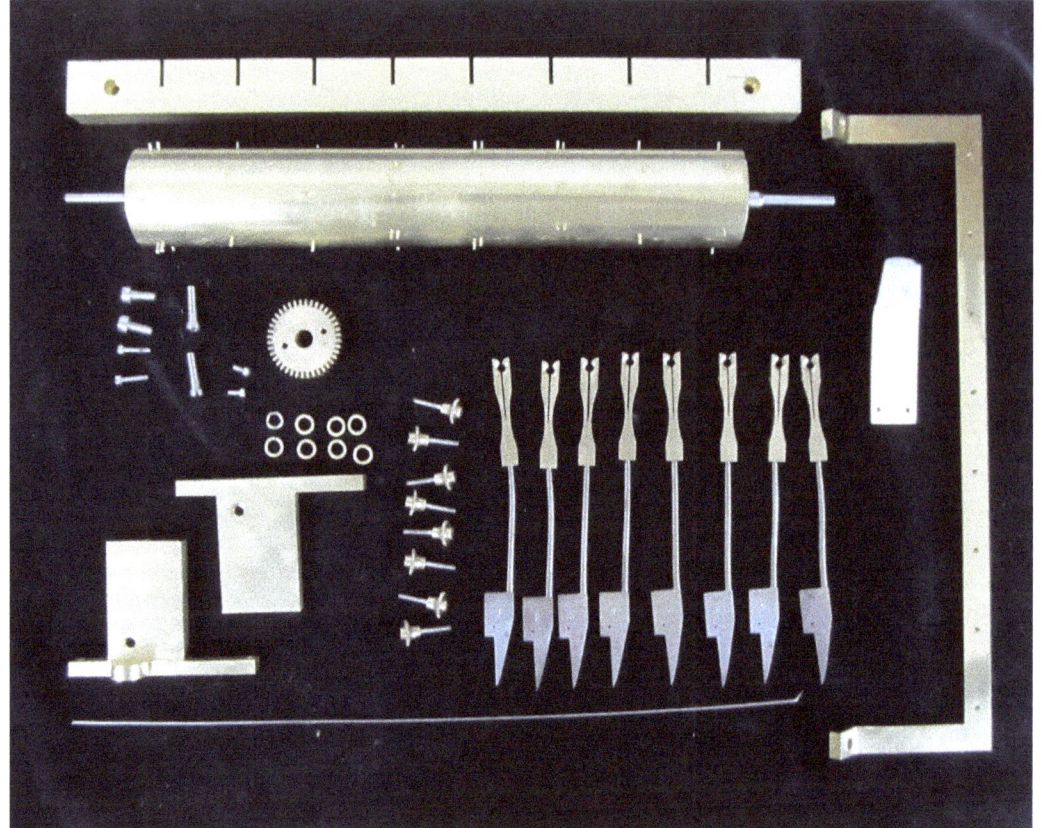

Figure 6.57. All of the individual parts of the cylinder assembly after they had been cleaned and restored and ready for final assembly.

Figure 6.58. Right view (chime train side) of the restored cylinder assembly mounted to the movement, showing the gearing required to transfer power from the movement to the chime assembly.

Figure 6.59. The black iron bracket used to hold the nine hanging tubes, the eight hammer spring leaves, and the eight (disassembled) striking hammers before they were cleaned. (See the finished assembly in Figures 6.63 and 6.64.)

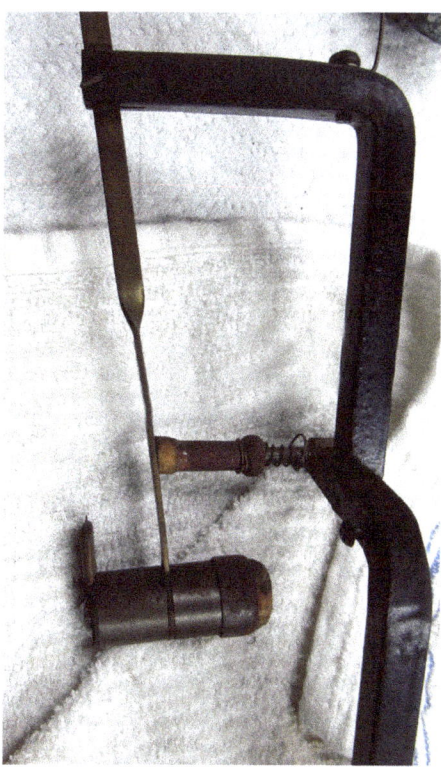

Figure 6.61. The hour tube striking hammer assembly before any restoration of the hammer and shock absorber assemblies.

Figure 6.60. The restored eight striking hammers with new leather pads on the right and the eight string adjustment clips on the left.

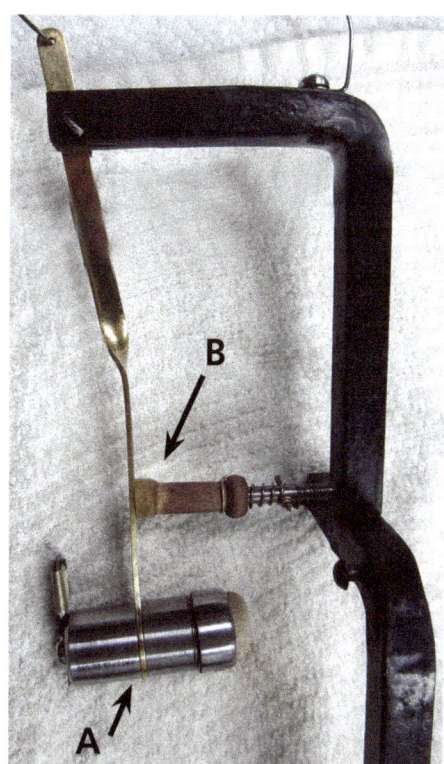

Figure 6.62. The restored assemblies after cleaning the hammer head, the hammer assembly "A," and shock absorber assembly "B."

Walter H. Durfee—Chapter 6 • **105**

Conclusion

This article was not written with the intent to tell someone how to restore a 9 tube tall clock but rather to illustrate and show the process and steps used by Dan Buffinga, a professional clock restorer. Over the years there have been many articles, lectures, discussions, and demonstrations on the different methods used in restoring case finishes, overhauling and cleaning movements, and re-silvering dials. Therefore, the restorer and I thought it would be best to leave the "how" to the individual, who was or would be, undertaking a similar restoration project. This neutral approach holds true to the material used in this restoration. So many commercial cleaning and restoration materials are on the market that we thought that the choice of which product to use should be left entirely up to the individual restorer.

Acknowledgments

I would like to give special acknowledgment to Dan Buffinga for supplying all of the photos that made this article possible and to Mr. and Mrs. Joel T. "Bo" Cheatham III, the owners of the Durfee Pattern No. 24 clock, for giving their permission to allow Buffinga to photograph his work and to allow the NAWCC to feature his clock in this article. I must also give special thanks to editor Therese Umerlik for her suggestion to turn this article into a photo essay and her assistance in guiding me through the entire process. The quality of the images and their layout would not have been possible without the excellent editing of associate editor Amy Klinedist. Thanks to my wife, Jo Burt, for her encouragement to take on the task and time to write this article. Special thanks to Raymond Beard for his suggestions and technical reviews.

Figure 6.63. The nine polished chrome tubes with new hanging cords are in the restored case.

Figure 6.64. Full view of the hanging tubes with the restored hammers and the new white hammer adjusting strings.

About the Restorer

Dan Buffinga started in the antiques business part time in 1995 while working full time as an automotive technician. Within a few years his interests became drawn more toward clocks. He began educating himself in correct and proper repair procedures, buying tools and equipment to do so. He started repairing and restoring clocks for others in 2005, working primarily on nineteenth-century French mantel clocks. In 2007, after a 30-year career as an automotive technician, he retired and became self-employed in the antiques clock business. He has been repairing and restoring clocks full time since and has specialized in fusee, chime tube, and musical clocks for the past five years. He joined the NAWCC in 2009. To learn more about his business, visit www.dlbantiques.com.

References and Notes

Chapter 1:

1.1. Elisha Durfee purchased the grist mill and the two adjacent properties on February 9, 1846, from William S. Thornton for the sum of $2,400. Terms of the mortgage agreement allowed Durfee to pay off the mortgage, with interest, in three payments: the first, one year from the purchase date; the second, 18 months from the purchase date; and the last payment, two years from purchase date. Dennis Jacobs, the owner of the original mortgage deed, gave us access to the deed.

1.2. Authors spoke with Hamilton Pease, who had conversed with Elisha Durfee.

Pease was born and raised in Providence, RI. He would pass the Durfee clock repair shop on his way home from school and would peek in the window at the clocks in the shop.

Finally he worked up enough courage to stop in and introduce himself to Elisha Durfee, and they immediately became friends. Pease later became a watch repairer and opened his own shop. Durfee then became his mentor on clock repair. Pease claimed that because of their friendship, the two spent many hours talking about clocks, watches, and their industry. Because of the difference in their ages, Pease believed that Durfee treated him like a son, much like Walter Durfee treated Elisha.

Pease was awarded the Fellow Award for his service to the National Association of Watch and Clock Collectors, Inc.

1.3. Authors spoke with Hamilton Pease, who had conversed with Elisha Durfee.

1.4. Authors spoke with Dana Blackwell, who had conversed with Elisha Durfee.

Blackwell (1917-2007) developed an interest in horology when he was a teenager in high school. He was educated at Munson Academy and Tufts University majoring in engineering. After graduation he taught mathematics, English, Latin and German for the next ten years.

Blackwell then spent the next 25 years engineering and designing aircraft precision instruments systems, obtaining several patents. However he continued to maintian his interest in horology.

After he left the aircraft industry, he became the vice president and chief engineer of E. Howard Clock and Watch Co., in Waltham, MA. Blackwell served as the curator and horological consultant for the American Clock and Watch Museum in Bristol, CT. He also served as a consultant for leading museums as well as articles on horology for US and international publications.

1.5. Authors spoke with Hamilton Pease, who had conversed with Elisha Durfee.

1.6. Authors spoke with Henry Fried, who had conversed with Elisha Durfee.

Fried (1918-1996) was an internationally renowned horologist. His career in horology covered teaching, industry consultant, expert witness, illustrator, lecturer, technical director, and author. He wrote 14 books, numerous pamphlets and hundreds of technical articles.

He learned watch repair skills from his father, a watch repairer in New York. After finishing school, he taught watch repair in New York City schools.

He was president of the New York City Horological Society and the New York State Watchmakers Association as well as vice president of the Horological Institute of America. He was the first American to receive the Silver Medal of the British Horological Institute.

He was honored with a Silver Star Fellow Award from the National Association of Watch and Clock Collectors. He also received an Outstanding Achievement Award from the United Horological Association of America. David Stout, "Henry B. Fried, 89, an Expert on Watches, Clocks and Time," Obituaries, *New York Times*, accessed on July 19, 2016. http://www.nytimes.com/1996/03/12/us/henry-b-fried-89-an-expert-on-watches-clocks-and-time.html.

1.7. This number will be discussed in depth in Part 2 and Part 3.

1.8. For a complete history of Durfee's girandole clocks, see "Walter H. Durfee & His Curtis Girandole Clocks," by Owen H. Burt, *NAWCC Bulletin*, No. 390 (February 2011): 4-12

1.9. Hershel Burt verified the movement when it came up for auction in later years.

Burt (1927-2000) was an early member of the National Association of Watch and Clock Collectors, Inc. along with his father Ed Burt, director of watch dial manufacturing at Waltham Watch Co. in Waltham, MA.

He later received the Fellow Award and Fellow Star Award for his contributions to horology. He served as a trustee and consultant to the Willard House and Clock Museum in North Grafton, MA, and the American Clock and Watch Museum in Bristol, CT. His expertise was in early American clocks and their makers.

1.10. There is an exception to this rule: Prior to 1887 a few 2-weight movements were not signed and in those cases the dial's nameplate was either signed "Durfee and Enches" or "Walter H. Durfee Providence."

1.11. Joseph Martines, "Contemporary Clockmaking: An Overview," *NAWCC Bulletin*, No. 191 (December 1977): 571-585.

Chapter 2:

2.1. Tran Duy Ly, *Long Case Clocks and Standing Regulators Part I: Machine Made Clocks*, Evanston, IL: Arlington Books, 1994.

2.2. In these chapters the word "manufactured" is interchangeable with the words "made" and "produced" since Walter H. Durfee never had an actual manufacturing plant or facility. Durfee purchased clock components from different suppliers and assembled them in his shop in Providence, RI.

2.3. Authors spoke with Hamilton Pease, who had conversed with Elisha Durfee. See endnote 1.2 for information about Pease.

2.4. Tom Spittler found this S. Harlow movement from Ashbourne, Derbyshire, England. Ashbourne is located approximately 55 miles southeast of Liverpool. This movement is found on page 393 in Volume 2 of John Robey's book titled *Longcase Clock Reference Book* (Derbyshire, UK: Mayfield Books, 2001).

Spittler retired as a lieutenant colonel from the US Air Force after 21 years of service. He holds degrees from Syracuse University in forestry and paper science and from the Air Force Institute of Technology in electrical engineering.

Spittler became interested in clocks while he was stationed in England. He has written several articles for the National Watch and Clock Collector, Inc.'s publication and *Clocks* magazine. He also coauthored a book on clockmakers.

He is considered an expert and authority on American and English grandfather clocks.

2.5. Authors spoke with Hamilton Pease, who had conversed with Elisha Durfee. See endnote 1.2 for information about Pease.

2.6. The brass dial is beautifully engraved "Vernon & Shepperd, Liverpool."

James Vernon and Thomas Shepperd worked as clockmakers in Liverpool: Vernon from 1754 to 1800 and Shepperd from 1734 to 1766. Therefore, we are using 1760 as the approximate date the clock was produced.

William H. Durfee sold the clock out of his antique shop to a local resident in Providence, RI; it has a perfect Durfee label like the one shown in Figure 2.4 .

The known provenance shows that the clock was owned by Theodore Collins, a professor at Brown University in Providence, RI, from the late 1800s until the mid-1930s. The clock remained in Providence until 2010. Whether Colliers purchased the clock directly from Durfee or was the second owner, we are not sure, but it appears the clock never left the Providence area until 2010.

2.7. The J. S. Jennens & Sons factory stamped "Walther H. Durfee, Providence" on the movement plate. We can assume that Walter H. Durfee & Co. made a special purchase order agreement with the manufacturer to purchase a large enough quantity of 2-weight movements to justify the factory-added expense of incorporating the added stamping operation. When Durfee stamped his name on the movement, he handstamped it on the top edge of the rear plate (Figure 2.48). Note that in the J. S. Jennens & Sons factory stamp in the center section, the stamp has two arrowheads in its design, similar to the arrowhead nameplate design that Durfee started to use in mid-1883 (Figure 2.11).

2.8. This information was confirmed by Tom Spittler, an American authority on the W. F. Evans & Sons Co.

2.9. These twisted columns are often referred to as barley twist columns. Note that the right twisted column twists in the oppostie direction of the left column.

2.10. Ly, *Long Case Clocks and Standing Regulators*, 146, Figure 342.

2.11. The Pattern 41 and Pattern 43 clocks in Ly's book titled *Long Case Clocks and Standing Regulators Part I: Machine Made Clocks*, Evanston, IL: Arlington Books, 1994.

2.12. We use the phrase "practically" because it appears that Durfee had a helper by the name of "Arnold" who took care of the shop while Durfee was in England, answered some of Durfee's correspondence, and serviced as his "clock winder" for his Providence clientele, until Elisha Durfee joined the company in 1914.

2.13. Authors spoke with Dana Blackwell, who had conversed with Elisha Durfee. See endnote 1.4 for information about Blackwell.

Chapter 3:

3.1. Authors spoke with Hamilton Pease, who had conversed with Elisha Durfee. See endnote 1.2 for information about Pease.

3.2. Authors spoke with Hamilton Pease, who had conversed with Elisha Durfee. See endnote 1.2 for information about Pease.

3.3. This Herschede's catalog can be found in Tran Duy Ly's *Long Case Clocks and Standing Regulators Part I: Machine Made Clocks*, Evanston, IL: Arlington Books, 1994, 148-154.

3.3. Authors spoke with Hamilton Pease and Dana Blackwell, who had conversed with Elisha Durfee. See endnote 1.2 for information about Pease and endnote 1.4 for information about Blackwell.

3.4. It's not likely true since we will learn that Durfee's cases were made in the United States after 1889.

3.5. Tran Duy Ly, *Long Case Clocks and Standing Regulators Part I: Machine Made Clocks*, Evanston, IL: Arlington Books, 1994.

3.6. Ibid., 145, Figure 339.

3.7. Ibid., 457, Figure 1061.

3.8. Ibid., 144-146.

Chapter 4:

4.1. Authors spoke with Dana Blackwell, who had conversed with Elisha Durfee. See endnote 1.4 for information about Blackwell.

4.2. Authors spoke with Henry Fried, who had conversed with Elisha Durfee. See endnote 1.6 for information about Fried.

4.3. Information on the J. J. Elliott & Co was obtained

from Tom Spittler's "J. J. Elliott and the American Hall Clock Industry" article that was published in the *NAWCC Bulletin*, No. 301 (April 1996). Much of his information came from Tran Duy Ly's book *Long Case Clocks* and the Herschede's 1885 trade catalog.

4.4. The word "tube" is interchangeable with the word "bell." Elliott and Harrington refer to their hanging tubes as bells. In this article the authors call their bells "tubes."

4.5. Tran Duy Ly, Longcase Clocks and Standing Regulators (Johnson City, TN: Arlington Books, 1994): 144, Figure 338.

4.6. Ibid., 145, Figure 339.

4.7. Ibid., 145, Figure 340.

4.8. Dials made using a design is laid out on a sheet of brass, the design openings are removed by hand sawing, and the remaining design pattern is then hand engraved. Making the dials requires hours of hand labor.

4.9. Engraved dials have their pattern or design laid out on a brass surface and an engraver hand engraves the design. These dials required much less labor than the pierced dial.

4.10. It strongly appears that Durfee paid a flat fee up front for the tube rights.

4.11. The United States Tubular Bell Company shared the building occupied by the Methuen Organ Company.

4.12. Durfee was not happy with Elliott's string adjustment of the tube chimes, because the string method was too hard and difficult to adjust after the clock was 10-20 years old. Therefore, in the 1920s he applied for a patent that would replace the string method, but there isn't any record showing that he ever used his patent rights.

4.13. Derek Roberts, *British Longcase Clocks* (Atglen, PA: Schiffer Publications, Ltd., 1990): 292.

4.14. Durfee Tubular Bells catalog, 1915.

Chapter 5:

5.1. J. N. Dunning (N for Nye) was erroneously documented as J. L. Dunning in this 1923 article by Durfee.

5.2. Walter H. Durfee, "The Clocks of Lemuel Curtis," *The Magazine Antiques* (December 1923): pp. 283-284.

5.3. Ibid., 183-184.

5.4. Lillian Baker Carlisle, "New Biographical Findings on Curtis and Dunning Girandole Clockmakers," *American Art Journal* (May 1978): p. 96.

5.5. Andrew H. Dervan, "Daniel Steele's Legacy," *Watch and Clock Bulletin*. Forthcoming.

5.6. Walter H. Durfee, "The Clocks of Lemuel Curtis," *The Magazine Antiques* (December 1923): p. 281, Figure 2. The figure caption states, "Curtis dials are iron, finely enameled and with the numerals delicately and beautifully applied. In the earlier examples Arabic numerals appear, in the later ones the Roman system is used. Owned by D. J. Steele."

Time is a gift. Embrace the present.
Join the NAWCC

Become a member today by applying at nawcc.org
or calling toll free 877.255.1849 (US and Canada) or 717.684.8261

Join the National Association of Watch and Clock Collectors, Inc., the largest international association dedicated to preserving and stimulating interest in horology, the art and science of time and timekeeping.

Our members are enthusiasts, students, educators, collectors, businesses, and professionals, who love learning about the clocks and watches they collect, preserve, and study.

Your dues support:
- World's leading research library on horology
- Largest public collection of timepieces in the Americas
- Educational programs that teach watchmaking and clockmaking skills as well as art and history of timekeeping

Your membership gives you:
- Six issues of the *Watch & Clock Bulletin*, an educational journal, and the *Mart & Highlights*, an advertising supplement and chapter magazine
- NAWCC's Library and Research Center's resources and assistance
- Webinars and workshops
- Online access to articles, videos, and archival materials
- Regional buying and selling venues
- Camaraderie at meetings, dinners, and events
- Free admission to the National Watch and Clock Museum in Columbia, PA
- Free or discounted admission to more than 250 museums and science centers

The National Association of Watch and Clock Collectors is a 501(c)(3) nonprofit organization.

www.ingramcontent.com/pod-product-compliance
Lightning Source LLC
Chambersburg PA
CBHW041411300426
44114CB00028B/2985